Astronomy Education, Volume 2
Best practices for online learning environments

AAS Editor in Chief

Ethan Vishniac, Johns Hopkins University, Maryland, USA

About the program:

AAS-IOP Astronomy ebooks is the official book program of the American Astronomical Society (AAS), and aims to share in depth the most fascinating areas of astronomy, astrophysics, solar physics and planetary science. The program includes publications in the following topics:

 GALAXIES AND COSMOLOGY

 INTERSTELLAR MATTER AND THE LOCAL UNIVERSE

 STARS AND STELLAR PHYSICS

 EDUCATION, OUTREACH, AND HERITAGE

 HIGH-ENERGY PHENOMENA AND FUNDAMENTAL PHYSICS

 THE SUN AND THE HELIOSPHERE

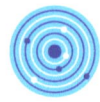 THE SOLAR SYSTEM, EXOPLANETS, AND ASTROBIOLOGY

 LABORATORY ASTROPHYSICS, INSTRUMENTATION, SOFTWARE, AND DATA

Books in the program range in level from short introductory texts on fast-moving areas, graduate and upper-level undergraduate textbooks, research monographs and practical handbooks.

For a complete list of published and forthcoming titles, please visit iopscience.org/books/aas.

About the American Astronomical Society

The American Astronomical Society (aas.org), established 1899, is the major organization of professional astronomers in North America. The membership (~7,000) also includes physicists, mathematicians, geologists, engineers and others whose research interests lie within the broad spectrum of subjects now comprising the contemporary astronomical sciences. The mission of the Society is to enhance and share humanity's scientific understanding of the universe.

Editorial Advisory Board

Steve Kawaler
Iowa State University, USA

Ethan Vishniac
Johns Hopkins University, USA

Dieter Hartmann
Clemson University, USA

Piet Martens
Georgia State University, USA

Dawn Gelino
NASA Exoplanet Science Institute, Caltech, USA

Joan Najita
National Optical Astronomy Observatory, USA

Bradley M. Peterson
The Ohio State University / Space Telescope Science Institute, USA

Scott Kenyon
Smithsonian Astrophysical Observatory, USA

Daniel Savin
Columbia University, USA

Stacy Palen
Weber State University, USA

Jason Barnes
University of Idaho, USA

James Cordes
Cornell University, USA

Astronomy Education, Volume 2
Best practices for online learning environments

Chris Impey and Matthew Wenger
Department of Astronomy, University of Arizona, Tucson, AZ 85721, USA

IOP Publishing, Bristol, UK

© IOP Publishing Ltd 2020

All rights reserved. No part of this publication may be reproduced, stored in a retrieval system or transmitted in any form or by any means, electronic, mechanical, photocopying, recording or otherwise, without the prior permission of the publisher, or as expressly permitted by law or under terms agreed with the appropriate rights organization. Multiple copying is permitted in accordance with the terms of licences issued by the Copyright Licensing Agency, the Copyright Clearance Centre and other reproduction rights organizations.

Permission to make use of IOP Publishing content other than as set out above may be sought at permissions@ioppublishing.org.

Chris Impey and Matthew Wenger have asserted their right to be identified as the authors of this work in accordance with sections 77 and 78 of the Copyright, Designs and Patents Act 1988.

ISBN 978-0-7503-1719-1 (ebook)
ISBN 978-0-7503-1717-7 (print)
ISBN 978-0-7503-1720-7 (myPrint)
ISBN 978-0-7503-1718-4 (mobi)

DOI 10.1088/2514-3433/abb3dd

Version: 20201201

AAS–IOP Astronomy
ISSN 2514-3433 (online)
ISSN 2515-141X (print)

British Library Cataloguing-in-Publication Data: A catalogue record for this book is available from the British Library.

Published by IOP Publishing, wholly owned by The Institute of Physics, London

IOP Publishing, Temple Circus, Temple Way, Bristol, BS1 6HG, UK

US Office: IOP Publishing, Inc., 190 North Independence Mall West, Suite 601, Philadelphia, PA 19106, USA

Contents

Preface	xii
Editor biographies	xiii
Contributors	xiv
Prologue: We All Are Online Astronomy Instructors	xviii

1 Guidelines for Teaching Astronomy Online — 1-1
Melody Buckner, Josie Strahle and Stephanie Tammen

1.1	Introduction	1-2
1.2	Flow of the Course	1-3
1.3	Interaction with Instructors and Peers	1-4
1.4	Engagement of the Student in the Learning Activities	1-4
1.5	Demonstration of Learning Outcomes by the Student	1-5
1.6	Conclusion	1-7
	References and Resources	1-7

2 Who Are We Teaching Online, and Why? — 2-1
Matthew Wenger

2.1	Introduction	2-1
2.2	Worldwide Online Education	2-2
2.3	Western Countries	2-3
2.4	Demographics and Motivations	2-4
2.5	Conclusion	2-6
	References	2-7

3 Effective Course Design — 3-1
Adrienne J Gauthier

3.1	Introduction	3-2
3.2	What is Your Teaching and Learning Philosophy?	3-4
3.3	Course Design Overview	3-5
	3.3.1 Course Design Methods	3-5
	3.3.2 Course Design Prep	3-8
3.4	Step 1: Developing Learning Objectives	3-9
	3.4.1 Big Ideas, Goals, and Questions	3-9

	3.4.2	Significant Learning	3-11
	3.4.3	Tutorial: Writing Learning Objectives	3-12
3.5	Step 2: Assessing Student Learning	3-18
	3.5.1	Types of Assessment	3-18
	3.5.2	Feedback Loops	3-20
	3.5.3	Assessment Strategies	3-21
	3.5.4	Designing Assessments that Align	3-22
3.6	Step 3: Creating Learning Experiences	3-25
	3.6.1	Active Learning Primer	3-25
	3.6.2	Choosing and Aligning Learning Activities	3-27
	3.6.3	Resources	3-28
3.7	Step 4: Putting It All Together	3-30
	3.7.1	Checking for Alignment	3-30
	3.7.2	Finding the Flow	3-31
	3.7.3	Communicating Your Course Plan	3-32
3.8	Resources for Teaching Online	3-32
3.9	Conclusion	3-33
	Acknowledgments	3-35
	References	3-35

4 Astronomy Education in Virtual Worlds and Virtual Reality — 4-1
Anthony Crider

4.1	Introduction	4-2
4.2	Astronomy Education in Virtual Worlds	4-2
	4.2.1	What Are Virtual Worlds?	4-2
	4.2.2	How Has Astronomy Education Been Done in Virtual Worlds?	4-3
	4.2.3	Getting Started in Virtual Worlds	4-8
4.3	Astronomy Education in Virtual Reality	4-10
	4.3.1	What Is a Virtual Reality?	4-10
	4.3.2	How Can Virtual Reality Be Used for Astronomy Education?	4-10
	4.3.3	Getting Started in Virtual Reality	4-13
4.4	Virtual Reality Virtual Worlds	4-14
	4.4.1	What Are Virtual Reality Virtual Worlds?	4-14
	4.4.2	Getting Started in Virtual Reality Virtual Worlds	4-16
4.5	Concluding Remarks	4-19
	References	4-19

5	**Massive Open Online Astronomy Courses**	5-1
	Paul Francis	
5.1	Introduction	5-1
5.2	MOOC Statistics	5-3
5.3	Why Teach a MOOC?	5-5
5.4	Which MOOC Platform?	5-7
5.5	What Topic Should I Teach?	5-8
5.6	Course Structure	5-9
5.7	Making the Videos	5-9
5.8	Assessment	5-13
5.9	Discussion Forum	5-16
5.10	Other Things to Think About	5-16
5.11	Conclusions	5-17
	References	5-17

6	**Using New Media and Social Media for Online Learning**	6-1
	Pamela L Gay	
6.1	Overview and History	6-1
6.2	Getting Started: A Practical Guide	6-3
6.3	Best Practices That Are Here to Stay	6-5
	6.3.1 New Media	6-5
	6.3.2 Social Media	6-6
6.4	Ethical Considerations	6-9
6.5	Conclusions	6-10
	References	6-10

7	**Education Through Exploration: A Model for Using Adaptive Learning to Teach Laboratory Science Online**	7-1
	Chris Mead, Ariel D Anbar, Lev B Horodyskyj and Donald Bratton III	
7.1	Introduction: What Problem Are We Solving?	7-1
7.2	The *What* and *Why* of Adaptive Learning	7-2
	7.2.1 What Student Needs Can Be Addressed?	7-4
	7.2.2 How Are These Needs Determined?	7-4
	7.2.3 How Is Adaptive Learning Delivered?	7-5
	7.2.4 What Does Adaptive Learning Look Like for Inquiry-based Science Labs?	7-6

7.3	The Education Through Exploration model	7-6
	7.3.1 Conveying Authentic Science	7-7
	7.3.2 Learning as a Journey	7-8
	7.3.3 Digital by Design	7-9
7.4	Examples of ETX Model and Adaptive Learning in Online Astronomy and Astrobiology Education	7-10
	7.4.1 Habitable Worlds	7-10
	7.4.2 BioBeyond	7-15
7.5	Implementing Adaptive Learning	7-17
7.6	Conclusion	7-20
	References	7-20

8 Key Online Resources for Teaching Astronomy — 8-1
Chris Impey and Andrew Fraknoi

8.1	Introduction	8-1
	8.1.1 Textbooks	8-3
	8.1.2 Laboratory Activities	8-4
	8.1.3 Compilations of Instructional Tools	8-4
	8.1.4 Online Homework and Learning Management Systems	8-5
	8.1.5 Concept Inventories	8-6
	8.1.6 Short Videos	8-7
	8.1.7 Interactive Tools	8-8
	8.1.8 Sky-viewing Tools	8-9
	8.1.9 Citizen Science	8-11
	8.1.10 Interdisciplinary Approaches to Teaching Astro 101	8-12
	8.1.11 Podcasts	8-13
	8.1.12 Image Collections	8-13
	8.1.13 Guides for New Instructors or TAs	8-14
	8.1.14 Databases of Astronomy Education Research Articles	8-15
	8.1.15 Blogs and Social Media Vehicles	8-16
	8.1.16 Miscellaneous Resources	8-16
	References	8-17

9 Epilogue: Lessons Learned from Transitioning to Online Learning During Spring 2020 During COVID — 9-1
Sanlyn Buxner, Nicole Gugliucci, Carl Ferkinhoff and Brian Jackson

9.1	Challenges and Considerations	9-1

9.2	Resources	9-2
9.3	Success Stories of Transition	9-3
	9.3.1 Winona State—Carl Ferkinhoff	9-3
	9.3.2 Saint Anselm College—Nicole Gugliucci	9-4
9.4	Final Thoughts	9-5

Preface

With online instruction becoming a modern necessity, many instructors are scrambling to develop courses with little prior knowledge about teaching a course online, and it can be challenging to know where to begin, and beyond that, how to develop a truly excellent and innovative online learning experience. This goal of this book is to help instructors teach astronomy courses in online learning environments. Whether you are a new instructor teaching for the first time, or an experienced teacher who faces the challenge of converting your in-person courses to online, the chapters in this book have been written to help you understand the opportunities and challenges as well as to provide information about resources and examples so you can build on the experience and expertise of your colleagues who have taught students virtually. Decades of experience and research on teaching and learning online have been gathered and summarized by the authors, all of whom are experts in instructional design, online instruction, education research, emerging technologies, and building communities of online learners.

The topics in this book have been carefully chosen in order to support instructors who are new to teaching online, while also providing information about more advanced topics that can help more experienced instructors improve their courses even further, and hopefully to spark new ideas for improving their online instruction. The topics in this volume include an introduction to teaching online, research-based information about online learners, a step-by-step guide on how to design effective online courses, astronomy instruction using virtual worlds, massive open online astronomy classes, using new and social media to build communities of online learners, the use of adaptive learning systems for online instruction, and key resources for teaching astronomy online. Each chapter is informed by research on learner-centered instruction, and many of the authors are active researchers in this area. All the chapters in this book have links to online instructional resources as well as references to research articles and other useful content. Since the ebook can be updated easily, the content will keep pace with any new innovations in online instruction.

Editor biographies

Chris Impey

Chris Impey is a University Distinguished Professor of Astronomy at the University of Arizona. He has over 180 refereed publications on observational cosmology, galaxies, and quasars, and his research has been supported by $20 million in NASA and NSF grants. He has won 11 teaching awards and has taught two online classes with over 180,000 enrolled and 2 million minutes of video lectures watched. Chris Impey is a past Vice President of the American Astronomical Society and he has been an NSF Distinguished Teaching Scholar, the Carnegie Council's Arizona Professor of the Year, and most recently, a Howard Hughes Medical Institute Professor. He's written over 50 popular articles on cosmology and astrobiology, two introductory textbooks, a novel called *Shadow World*, and eight popular science books: *The Living Cosmos, How It Ends, Talking About Life, How It Began, Dreams of Other Worlds, Humble Before the Void, Beyond: The Future of Space Travel*, and *Einstein's Monsters: The Life and Times of Black Holes*.

Matthew Wenger

Matthew Wenger is an Education Program Manager with Steward Observatory at the University of Arizona. He develops and teaches online classes and has developed massive open online courses (MOOCs) for Coursera and Udemy that currently have over 137,000 students enrolled. In addition to online instruction, Matthew conducts and publishes educational research and manages online education and outreach programs for public audiences, including the YouTube channel "Active Galactic." He previously served as a postdoctoral researcher in free-choice learning at Oregon State University and has more than 20 years of experience with science education in both informal and formal learning environments.

Contributors

Ariel Anbar is the Director and Founder of the Center for Education Through Exploration (ETX) at Arizona State University as well as an ASU President's Professor in the School of Earth and Space Exploration and the School of Molecular Sciences, and a Howard Hughes Medical Institute Professor. He is an active Researcher and Educator with broad interests in the past, present, and future of Earth as an inhabited world, and the prospects for life beyond. His research and much of his teaching center on the fields of environmental and isotope geochemistry, Earth history, and astrobiology. Ariel's passion for science combined with his desire to transform how students learn inspired the Habitable Worlds and BioBeyond courses and later catalyzed the formation of the ETX Center.

Donald Bratton leads the learning design team at the Center for Education through Exploration at Arizona State University. Don's work has ranged from full university-level courses to innovative single lesson packages, focused on finding new ways to combine active learning and adaptive courseware to help students learn. Don is a veteran science educator and a National Board Certified Teacher who has taught Chemistry and Biology to students from sixth grade through college. He earned a BS in Chemistry from the University of Southern Mississippi and an MS in the same from Florida State University with concentrations in Biochemistry.

Melody Buckner has been the Associate Vice Provost of Digital Learning and Online Initiatives for the University of Arizona since 2019 July. She served as the Interim Dean of UA South (the branch campus and the College of Applied Science and Technology for the University of Arizona) from 2016 to 2019. From 2014 to 2019 she served as the Senior Director of Digital Learning and Online Education at the University of Arizona. Melody created an office of instructional design, visual design, video production and quality assurance for the Arizona Online. Before coming to the University, she served as an Instructional Designer in Professional Development and as an adjunct faculty for Pima Community College. She conducts research and teaching in the areas of educational technology, digital and creative literacies and multicultural education.

Sanlyn Buxner is an Associate Research Professor at the University of Arizona and a Senior Research Scientist and Senior Education and Communication Specialist at the Planetary Science Institute. She teaches undergraduate and graduate science and science education research courses. Her research investigates science literacy, quantitative literacy, and motivations of learners in formal and informal science courses as well as the impacts of science research experiences for undergraduates and K-12 teachers.

Anthony Crider is a Professor of Astrophysics at Elon University and the current Chair of the Board of the Reacting Consortium, a non-profit dedicated to developing role-playing games for higher education. His interest in astronomy visualization led him to create virtual planetariums, telescopes, and lunar landscapes within the 3D online world of Second Life and eventually to co-found the SciLands, a mix of universities, government agencies, and others dedicated to science education and outreach in virtual worlds. Back in the "real" world, he uses Reacting to the Past role-playing games in his classes, including his own game, The Pluto Debate: The International Union Defines a Planet, which was the first of many science reacting games to receive funding from the National Science Foundation. Dr Crider has professionally written and spoken on a variety topics at the intersection of astronomy, education, visuals, and games, including the morphologies of active galaxies, the teaching of visual literacy in astronomy, the changing popularity of social virtual worlds, and the use of "epic finales" as a form of experiential assessment in SETI-themed and other classes.

Carl Ferkinhoff is an Associate Professor of Physics at Winona State University, in Winona Minnesota. He is program lead for the physics teacher education program. His responsibilities include teaching the 5–12 Science Methods course for all future middle and high school science teachers at Winona State. In his own teaching he implements both research validated instructional techniques and assessments to assess both student learning and his own abilities as a teacher. His research involves studying galaxies in the early Universe at submillimeter wavelengths and developing instrumentation that enables these observations. This work is driven largely by undergraduate students in Winona. Prior to moving to WSU, Carl completed a postdoc at the Max Planck Institute for Astronomy in Heidelberg Germany, his PhD in Astronomy at Cornell University, and taught high school Physics & Chemistry in Baltimore, MD through Teach For America.

Andrew Fraknoi retired from full time college teaching in 2017, and now teaches short courses for retired people at the University of San Francisco and San Francisco State University. He served as Executive Director of the Astronomical Society of the Pacific from 1978 to 1992, where he created a number of national astronomy education projects, programs, and publications. Later, he was founding co-editor of the journal *Astronomy Education Review*. He is lead author of the free, open-source astronomy textbook, *Astronomy*, published by the non-profit OpenStax project at Rice University, and edited the resource guide, *The Universe at Your Fingertips 2.0*. One of his hobbies is collecting music inspired by serious astronomical ideas. For more, see: www.fraknoi.com.

Paul Francis is an Astrophysicist and Educator. After doing a PhD at the University of Cambridge, he worked at the University of Arizona and the University of Melbourne before settling at the Australian National University. His research centers of quasars, high redshift galaxies, and novel teaching techniques. He has won many national awards for teaching and public outreach.

Adrienne Gauthier is a Learning Designer at Dartmouth College. She collaborates with STEM faculty through course design projects that focus heavily on learner-centered strategies. Other projects include managing the undergraduate Learning Fellows Program, designing MOOCs, and investigating new educational technologies. Previously, she worked as an Instructional Technologist and Astronomy Educator in the Department of Astronomy at the University of Arizona, working with faculty and other astronomy community professionals to innovate with technology-enhanced learning. These included student creative projects and IYA 2009 in Second Life, co-designing online and hybrid astronomy courses, collaborating on the Astronomy Visualization Metadata standard, and managing open online astronomy resources like Astropedia, now known as TeachAstronomy.com.

Pamela Gay is an Astronomer, Techhead, and Creative focused on using new media to engage people in science and technology. Through CosmoQuest.org, she works to engage people in both learning and doing science. She is also cohost of the award-winning podcast, Astronomy Cast. Through this weekly podcast, Fraser Cain and Dr Gay take listeners on a facts-based journey through our Cosmos, exploring not only what we know, but how we know it. In 2018, she was inducted into the Podcasting Hall of Fame. Professionally, she is a senior scientist at the Planetary Science Institute and personally she lives on the Internet as StarStryder on most major platforms.

Nicole Gugliucci is an Astronomer and Education Researcher with a background in radio astronomy instrumentation and a passion for sharing the universe. She is an Assistant Professor of Physics at Saint Anselm College in Manchester, New Hampshire. In addition to teaching introductory and advanced physics and astronomy, Dr Gugliucci gives talks on campus and with local groups about radio astronomy, citizen science, and the convergence of science and science fiction. At night, she can often be found leading constellation tours and telescope observing sessions at the college's observatory. Her current work focuses on recruiting secondary education teachers from the ranks of the college's STEM majors through authentic teaching experiences. She also teaches science to pre-service elementary teachers and studies the motivations of citizen scientists.

Lev Horodyskyj is an Educational Innovator at Science Voices and Blue Marble Space Institute of Science. He led the design and development of Habitable Worlds and continues to work on curricular materials spanning topics from astronomy to sustainability for delivery in various digital and hybrid formats. He has developed various simulators that help students visualize complex processes, from star formation to the carbonate–silicate cycle to cellular metabolism. Lev also analyzes the significant amounts of data generated through digital systems to investigate the effectiveness of these curricula and identify technological and pedagogical gaps that still need to be addressed. Lev holds a PhD in Geosciences and Astrobiology from Pennsylvania State University.

Brian Jackson is an Associate Professor in the Physics Department at Boise State University. Before joining the faculty at Boise State, he was a postdoctoral fellow at the Carnegie Institution of Washington's Department of Terrestrial Magnetism in Washington DC and, before that, at NASA's Goddard Space Flight Center in Greenbelt MD. He earned his PhD in Planetary Science from the University of Arizona's Lunar and Planetary Laboratory in Tucson AZ, and BS in Physics from Georgia Tech in Atlanta GA. His research focuses on exoplanets and planetary aeolian processes, and his teaching centers on introductory astronomy and advanced undergraduate courses.

Chris Mead is an Assistant Research Scientist working in the Center for Education Through Exploration (ETX) at Arizona State University. He is a discipline-based education researcher, having earned his doctorate in isotope geochemistry and geoscience education at ASU. His current work includes both research and evaluation projects looking at how digital technologies can improve science education—whether in-person or fully online.

Josie Strahle has a BS in Biology and an MA in Teaching & Teacher Education with an emphasis in science from the University of Arizona. She serves as a Senior Instructional Designer for Digital Learning at the University of Arizona, focusing on the development of fully online STEM courses. Before beginning her instructional design career, she taught high school science for seven years and continues to apply lessons learned about engagement in science courses to her current work. She loves to explore how technology can be used to improve instruction and bolster teaching and learning experiences.

Stephanie Tammen earned her BS in Nutrition Science at the University of Arizona before moving to Boston to complete a PhD in Biochemical and Molecular Nutrition at Tufts University. Upon graduation, Stephanie transitioned into education via a postdoctoral fellowship in curriculum design and evaluation. She then co-directed an online graduate-level program that taught biomedical science to high school instructors nationwide at Tufts University School of Medicine. She also worked with high school teachers to research how partnerships between instructors and biomedical scientists may improve biology courses and student outcomes. Stephanie currently is an Instructional Designer for Digital Learning at the University of Arizona, where her primary focus is bringing STEM courses online.

Prologue: We All Are Online Astronomy Instructors

In the Spring of 2020, educators around the world were faced with a seemingly impossible task: move entirely to remote teaching and do it FAST. In North America, the COVID-19 pandemic became an issue for universities and K-12 schools in March, and for many the normal spring break holiday became a time of transition from classroom to fully-online instruction. As anyone who has tried to plan a course knows, one stress-filled week is not nearly enough, and the resulting chaos had instructors of all levels scrambling to define their own way forward, using a myriad of different online content management systems, video streaming platforms, and recorded videos. This was not ideal; it was an emergency transition that required all of us to adapt in real-time to emerging challenges related to everyone—students, faculty, and TAs—losing access to campus resources, ranging from lab equipment to computers to safe housing. As instructors, we can't solve all our students' problems, but we can make their lives easier through instruction that takes advantage of best practices in online instruction.

This prologue is being written as we enter a new school year that, for most of us, will begin with classes fully online, and mixed face-to-face and online instruction. As we have the time to take a more considered approach to this new year, and future years, we hope this book will help you find a smoother path forward as we all learn to teach online in more effective ways.

For this year and future years, at least we have time to prepare our courses from the beginning, although we are often being asked to do the impossible: to prepare courses that can be delivered online, in-person, or both, and meet the needs of students who may choose to be in class and not in class at different times. For the foreseeable future, we need to be prepared for everything to be virtually presented for at least some of our students at any time. We need to figure out how to build labs, observing, and collaboration components into our mixed method classes. Chapters from this volume, as well as Volume 1, will be valuable as we prepare for these new challenges. Although most of us did not set out to be online astronomy instructors, we now have the opportunity to improve in this arena and integrate effective teaching practices and tools for students who may not be ready to be online students.

The chapters in this book were written prior to COVID-19, and therefore do not reflect solutions for all the problems we will face. At the end of this book, we include an epilogue of additional lessons learned by instructors at diverse institutions, ranging from small private colleges to massive public universities. While everyone's experience is unique, we found that no matter the size of the institution of higher education, many of the challenges experienced by students and faculty were similar. The rapid transition to online learning challenged instructors who had not used their course management software heavily with students before, and created a steep learning curve for instructors and students who were unprepared for such an extreme shift.

For instructors who had not seen students due to spring break, communications about changes of course delivery and assignments became a priority. Student confusion was a constant theme as each instructor at a given institution was given latitude about how they wanted to complete the semester. At most institutions, students were given the choice to switch from regular course grades to pass/fail grades, sometimes decreasing motivation for participating in the second half of the semester. These problems are ones we now have the ability to plan for, and to define solutions for.

While there have always been students ready to jump eagerly into online learning opportunities, this isn't true of the majority of on-campus students. Although many of our students might be digital natives, they may not be digitally fluent. For instance, as researchers we likely know how to take screenshots of our computers quickly, but our students may not. As we build our courses, we need to build in time to show students how to use the tools that we might take for granted, teaching them skills such as how to scan documents to PDF with their phones, or a myriad of other skills they will need to submit work that was once turned in on paper.

Beyond these instructional difficulties, during the spring 2020 semester we also experienced more personal challenges. Throughout this first chaotic semester, we all became aware of the incredibly unusual circumstances under which we were (and still are) living and had to be cognizant of many more external factors; our traditional students instantly had other priorities such as family, jobs, and their physical and mental health to contend with. All of these became even harder for non-traditional students. Courses take a lower priority as students deal with caring for their families, having to move off campus or out of the state and country, and the stress of having their entire world turned upside down. Given the circumstances, our astronomy courses, especially non-major courses, were not among students' most pressing priorities.

The authors of this text don't have the required expertise to guide you in supporting students through all these struggles, but we would remind you that no one can learn effectively when they are in an unstable living situation, and as social beings, research shows we learn better through interactions; interactions that are now so much harder for so many people. As you plan your semester, we would encourage you to build compassion into your curriculum.

Without giving up on the semester or student learning, we learned we must step back and review the key learning goals for our courses as well as those mandated by our college and university requirements. During the spring 2020 semester, we found many things in a course can be cut while still achieving core learning goals. It isn't easy, but it can be done, and the astronomy education community has supported one another in finding new ways forward. The coming school years will not be easy, but quality education can still happen. Now more than ever, we think all of us can appreciate the ability of astronomy to take us out of this world and carry us into the stars.

Sanlyn Buxner
Nicole Gugliucci
Carl Ferkinhoff
Brian Jackson

Astronomy Education, Volume 2
Best practices for online learning environments
Chris Impey and Matthew Wenger

Chapter 1

Guidelines for Teaching Astronomy Online

Melody Buckner, Josie Strahle and Stephanie Tammen

Although teaching online is becoming more common, there are a lot of issues to consider when moving from a face-to-face classroom to a fully online learning environment. Some of these issues may not be obvious to instructors new to this modality of teaching. In this chapter, we provide an overview of considerations instructors need to take into account as they develop their online class and instructional materials. The material has been organized into four major areas for instructors to consider: (1) the flow of the course, which includes design, pedagogy, methodology, and technology, (2) how to facilitate interactions with the instructor and peers, (3) engaging students in the learning activities, and (4) demonstration of learning outcomes by the students. This chapter introduces instructors to these crucial concepts and provides resources, as well as connections to later chapters in this volume. Our goal is to help instructors orient themselves as they develop content, design environments, and prepare to teach online.

Chapter Outcomes

By the end of this chapter, readers will be able to:
- Describe the importance of course design and flow for the success of online classes.
- Describe best practices for instructor presence and successful feedback.
- Give examples of low-stakes assessment and techniques for scaffolding learning.
- Describe how automation and multimodal techniques can improve online teaching.
- Describe pathways for student agency in demonstration of learning.
- Find resources to develop capstone experiences for authentic application of learning.
- Evaluate their current courses and plan for improvement based on these guidelines.

1.1 Introduction

Teaching online is both a similar and a drastically different endeavor from teaching students in a face-to-face classroom. Thankfully, over the past two decades higher education instructors, education researchers, faculty developers, and learning designers have implemented and tested a variety of strategies to figure out what can work best. It is not just about effective and aligned course design, it is also about the learning environment that you create and foster. Some key questions to consider as you think about course design are:

- How can you promote inclusivity and a sense of belonging?
- What strategies will help make your online course inviting, engaging, and feel like a community?
- How will you get to know your students as people, and help them get to know (and help) each other?

Chickering & Gammon (1987) share the seven principles for good practice in undergraduate education and a second paper describes their implementation in technology enhanced education (Chickering & Ehrmann 1996). These principles are: (1) encourage student–faculty contact; (2) encourage cooperation among students; (3) encourage active learning; (4) give prompt feedback; (5) emphasize time on task; (6) communicate high expectations; (7) respect diverse talents and ways of learning. A variety of authors have written useful articles about the practical application of these principles, including Gorsky & Blau (2009).

With proper planning and the right tools, teaching science online can be as natural as teaching in the classroom. Online learning environments provide many opportunities and pose a few challenges that we will address in this chapter. There are four areas of importance to keep in mind when teaching an online course:

1. Flow of the course—includes design, pedagogy, methodology, and technology.
2. Interaction with instructor and peers.
3. Engagement of the student in the learning activities.
4. Demonstration of learning outcomes by the student.

There is a common concern that online courses can be isolating and dull. In a quality online course, however, instructors often find they know their online students better than their in-person students because they can easily monitor progress and are in more constant communication with them. What makes a quality online course? To answer this question, we will provide some guiding principles and methods for good online course design and teaching. The first guiding principle is:

When transitioning to an online course, think about the methodology or pedagogy before engaging in the technology.

It is important for an online course to have a good design, otherwise the students will become frustrated, and with no one to guide them, they may miss steps in the learning activities or give up altogether. As an example, one would not start building a house without detailed plans. The same is true for teaching online. For more details about the process of course design, Chapter 3 of this volume is designed as a "workshop in a box" that will guide you through the process of designing an online course. If your institution has the resources, you may also be able to work with an instructional designer who can make suggestions or review course materials. In addition, there are a number of online resources, such as Quality Matters (https://www.qualitymatters.org/) which can provide additional course design and course evaluation resources. Finally, remember to look at other online courses for ideas that can be used in the construction of a new course.

1.2 Flow of the Course

When considering the design of an online course, an instructor should think about the outcomes and the kinds of interactions they want to see students engage in. Let us go back to the house example, when you are planning your design, a good way to start is to think about the parties you want to eventually host in your home. For a successful party the host wants the guest to flow through the house and engage with others for an enjoyable experience. The same is true for an online course. How does a science instructor go about throwing an online party where students learn the elements of science, engage with each other, and implement scientific principles in a way that shows they have learned? Some of this can be accomplished by the visual layout and the organization of the course. The instructor wants students to easily navigate the online space and engage with one another throughout the adventure of learning the content. This brings us to our second guiding principle:

Provide a clear progression through the learning activities, and a clear and repeating structure for each unit that is consistent throughout the course.

There must be a clear beginning, and it should be immediately understandable where students should start. A document or a lesson titled "Start Here" is a great way to reduce ambiguity. When a student enters an online course for the first time, it is important to engage them or even excite them with the content they will be exploring in the course. This is a perfect opportunity to use the art of storytelling. The online environment allows media to assist in this engagement. There are many videos on the Internet that can be utilized if time is of the essence, but the best practice is for the instructor to introduce themselves and talk about the course. It is important to bring the "why am I here" and help the student make a connection to the person who will be guiding them through the content.

A clear structure and an engaging introduction will also allow students to navigate the course more autonomously with less micromanaging from the instructor. This brings us to our next guiding principle:

Build in as much automation as possible to the online course.

Taking the time to automate things like lessons and feedback gives the instructor more opportunity to engage with students on a higher level and individualized questions. The instructor can concentrate on individual communication and facilitation of higher-level assignments. For example, a physics course might have automated lessons on basic concepts that are designed to catch common misunderstandings and pitfalls, leaving the instructor with more time to assist students with an experimental design project. Such detailed development takes time (another reason to work with an instructional designer), but the payoff is worth it.

1.3 Interaction with Instructors and Peers

Although best practice is varied, an online instructor should be in the course and responding to students at least two to three times a week at minimum. It is important to communicate up front how often students will hear from you and the time it will take for you to respond to a personal request. This leads us to our next guideline:

Leverage online communication tools to connect with students synchronously and asynchronously.

Try experimenting with communication and feedback methods that go beyond text: provide feedback via audio or video message, post video announcements instead of text announcements, and provide office hours on a virtual meeting platform. Examples include video and audio tools, virtual meeting tools, discussion board, chat tools, Twitter, and texting apps. (Skype or Zoom for virtual meeting and VoiceThread for visual discussions are a few suggested applications.)

1.4 Engagement of the Student in the Learning Activities

Students bring their own backgrounds, misconceptions, and knowledge to the classroom. This means that what helps one student learn, may not be helpful for everyone. In comparison to traditional face-to-face courses, online learning excels at easily providing multiple techniques to teach a scientific concept. Our next guideline:

Use multimodal techniques for teaching scientific concepts.

This includes visualization, animations, and interactive simulations along with video explanations. Examples for Astronomy 101 might include, The Big Dipper in 3D visualization, a Proton to Proton animation or a Lookback Time simulator. For more ideas on how to use simulations interactively in the introductory astronomy classroom, see Chapter 5 in Volume 1 in this book series titled "Evidence based instruction for introductory courses" (Impey & Buxner 2019). Presenting materials in a variety of ways allows students to learn the content in a way that makes sense to them. Instructors can post readings, recorded online lectures, visualizations, animations, and interactive simulations. It is important to note that recorded online lectures should be kept short (under ten minutes) and be created specifically for teaching content online. These videos not only build a connection between the instructor and students but allow students to return to the lecture or parts of the lecture to reinforce difficult concepts and topics that need clarification. Visualizations and interactive simulations allow students to see the content in a manner that can be missing from a face-to-face course. For example, microscopic processes can be easily visualized, and using a simulation, systems can be manipulated, and effects instantly seen (e.g., changing the mass or length of a pendulum, changing the pressure or volume of a gas).

In addition to multimodal engagement, instructors can select assignments that have real-world meaning for students. Our next guideline is:

Utilize authentic (or real-world) assessments such as research or citizen science projects.

Not only do these assignments relate to real-world problems, they also promote digital literacy. Authentic assessments encourage students to learn how to use technology to present data, communicate, and interact in the digital world, which are important skills in today's work environment. Examples include a citizen science project or a laboratory exercise. See the list of resources below for links to resources and examples. For additional resources, there are two chapters in Volume 1 of this education series that contain helpful information about how to incorporate real-world science into introductory astronomy classrooms. Chapter 7 discusses "Authentic Research Experiences in Astronomy to Teach the Process of Science" and Chapter 8 is about "Citizen Science in Astronomy Education".

Teaching science students to thrive in a digital world includes teaching digital literacy skills, which is the ability to use technologies to identify, access, manage, integrate, evaluate, analyze, create, and communicate information. For the online science course, the student is already in a digital environment allowing them to easily access technology for mining data sets or sharing their own data. Through the guidance of an online instructor, digital literacy skills specific to their field can be modeled, explained and implemented for high impact results. An application for promoting this is an authentic capstone project. For example, students might use specialized software to analyze spectral data to determine the chemical composition of a star or other astronomical object. In addition to the chapters in Volume 1 mentioned above, see the list of resources at the end of this chapter for a link to John Dutton's astronomy class at Penn State that uses such a capstone project. Results could be visualized and communicated via an online scientific poster and online presentation.

1.5 Demonstration of Learning Outcomes by the Student

In isolation, reading or watching lectures are generally not effective approaches to retaining knowledge. Science courses pose unique learning challenges with complex nomenclature and potentially intimidating material. This brings us to our next guideline:

Include low-stakes assessments in each lesson.

Although these are a way for students to demonstrate their learning, low-stakes assessments are more for self-reflection than for demonstrating understanding to an instructor. Low-stakes assessments assist with scaffolding of complex topics and provide students the opportunity to be wrong in a safe learning environment. Luckily, the online environment is ideal for low-stakes assessments, or opportunities to test one's knowledge independent of grading. Beyond improving retention, low-stakes assessments (1) allow students the opportunity to practice and fail, (2) give students the chance to understand their own thinking and monitor their own progress, and (3) reinforce key learning outcomes and prepare students for graded assignments. Examples include questions embedded in readings or videos, repeatable quizzes, or lessons built in interactive authoring tools. By leveraging the associated analytics, instructors can more readily keep tabs on student progress (a significant benefit, as struggling students can easily slip through the cracks in large science courses).

Building on low-stakes assessments, when a fuller capacity of the online platform is harnessed, a course also contains pre-built feedback and adaptive learning pathways. Each time a quiz question or knowledge check is written, provide follow-up feedback that is specific to the answer choice. This might entail correcting a misconception that an incorrect answer elucidates, directing students to content that relates to the question, or expanding upon a correct answer. With extra development time, adaptive learning pathways can be constructed, branching at points that depend on answers to questions. In this way, students needing extra help/time on any given concept can receive it, and students who have a firmer understanding are guided onward.

A common question about teaching science online is how to meet learning outcomes typically obtained in physical lab spaces. This brings us to our next guideline:

Reevaluate the learning outcomes of on campus labs to determine if they can be achieved using existing data sets, a virtual lab simulation, or a home lab kit.

The learning outcomes of the lab should first be reviewed and evaluated in terms of the value of the physical hands-on experience. Many times, the learning outcomes can be achieved without traditional hands-on labs. For example, if the learning outcome is for students to develop and test a hypothesis, this can be achieved by using existing data sets, a virtual lab simulation, or a home lab kit. These are commercial or DIY kits created specifically for labs to be conducted at home. Open educational resources for ideas in this area are also available. Information about two resources, one unique to astronomy and the other to general teaching online with lab work can be found in the list of resources at the end of this chapter. The final guideline comes from best practices in teaching and learning for face-to-face instruction, however, technology tools used in online environments may be particularly well suited to implementation:

Provide several pathways for students to have agency in how they access and learn content and how they want to demonstrate learning outcomes.

Examples include letting the student decide if they want to write a paper, create an infographic, or even produce a media piece. Students could potentially record videos separately in a group project, then edit them together for the final product. As mentioned above, this is not unique to online classes, however, the digital technologies already being used in these environments make the options and implementation even easier and more interesting. If designed well, all the assignments can be graded with the same rubric for assessing the learning outcomes. Not only do students get to decide when and where they can access their learning experience, but they have some ability to control the process for completing their work. This is turn creates a more meaningful and deeper learning experience for students. Instructors can create environments where agency reinforces the learning experience, giving students the ability to discover independently and not just forcing them into a meaningless task for the sake of the grade (Darby & Lang 2019, pp. 159–160).

Even more than in face-to-face courses, it is important to make expectations for assignments clear to the students. The next guideline is:

Use grading rubrics whenever possible.

A comprehensive grading rubric is an excellent proactive feedback tool that allows students to know at the beginning of an assignment how they will be graded. Rubrics are also good to level the playing field for subjective grading. This keeps the instructor focused on the learning outcome for the assignment. Providing examples of student work can also be beneficial scaffolding tools (it is best to give varied examples to prevent copying).

1.6 Conclusion

Online teaching and learning have become a vital part of the educational landscape. What was once outside of the traditional format of education has become a mainstream form of delivery in today's courses. Although many of the teaching strategies used in face-to-face classes can be useful for developing an online class, there are significant differences that can make adapting to an online learning environment challenging. As instructors begin the process of developing their new courses, or to move an existing class online, social factors and interactions with and between students need to be explicitly considered. There are many innovative ways to communicate with students besides relying on in-person communication and existing classroom norms to regulate student participation. Some areas instructors need to consider when designing an online course are how to promote inclusivity and a sense of belonging, how to make online courses inviting, engaging, and feel like a community, how to engage with students and facilitate way they engage with each other and the content. Online classes open up new opportunities for instructors to use instructional technology for communicating, learning, and assessing students. Many of these are new to instructors who are only in the face-to-face classroom setting. We hope that this chapter has given you some new ideas and insights for beginning to develop your online courses.

References and Resources

Chickering, A. W., & Gamson, Z. F. 1987, AAHE Bull, 3, 7
Chickering, A. W., & Ehrmann, S. C. 1996, AAHE Bull, 49, 3
Darby, F., & Lang, J. M. 2019, Small Teaching Online: Applying Learning Science in Online Classes (New York: Wiley)
Darby, F. 2019, Small Teaching Online: Applying Learning Science in Online Classes, https://teachinginhighered.com/podcast/small-teaching-online/, https://lacol.net/small-teaching-online/
Gorsky, P., & Blau, I. 2009, Int Rev Res Open Distrib Learn, 10, https://doi.org/10.19173/irrodl.v10i3.712
Impey, C., & Buxner, S. 2019, Astronomy Education Volume 1: Evidence-based Instruction for Introductory Courses (Bristol: IOP Publishing)
OLC Session 2018, Teaching Science Online: Learning from the Experts, https://onlinelearningconsortium.org/webinar/teaching-science-online-learn-from-the-experts/

The resources below have been intentionally gathered to give you multiple perspectives on online teaching. Whether it's a small undergraduate course or a world-wide massively open online course (MOOC), you should be able to find strategies in the resources below that will work with your teaching philosophy.

General Resources for Teaching Online:
- Flower Darby's lens on James Lang's *Small Teaching: Everyday Lessons From the Science of Learning* is summarized in the Inside Higher Ed article, "Small Teaching Online" (https://www.insidehighered.com/digital-learning/article/2019/06/26/bringing-small-teaching-online-classroom). Read the interview and then get their new book Small Teaching Online: Applying Learning Science in Online Classes (Darby & Lang 2019).
- Review strategies from Michelle Pacansky-Brock for How to Humanize Your Online Course (https://brocansky.com/humanizing-infographic) and help your students be engaged, feel valued, and create community. What is meant by "humanizing"? Check out this EdSurge article by the same author, How to Keep the Human Element in Online Classes (https://www.edsurge.com/news/2016-04-27-how-to-keep-the-human-element-in-online-classes).
- The Innovative Learning team at Purdue University has put together an in-depth guide to online teaching and learning. Check out their PoRTAL (Purdue Repository for online Teaching and Learning) resource (https://www.purdue.edu/innovativelearning/supporting-instruction/portal/).
- In How to Be a Better Online Teacher ADVICE GUIDE (https://www.oercommons.org/authoring/18513-a-compilation-of-free-laboratory-activities-for-as/view) (Darby) in the Chronicle of Higher Education, you will find "advice on how to make your online pedagogy as effective and satisfying as the in-person version" (Darby 2019).
- UC Davis' open resource can help you think about Being Present in Your Online Course (https://canvas.ucdavis.edu/courses/34528/pages/being-present-in-your-online-course).
- Melissa Wehler shares Five Ways to Build Community in Online Classrooms (https://www.facultyfocus.com/articles/online-education/five-ways-to-build-community-in-online-classrooms/) through the Online Education section of Faculty Focus (https://www.facultyfocus.com/topic/articles/online-education/).

Resources for Teaching Science Labs Online:
- Fraknoi, A. 2018, A Compilation of Free Laboratory Activities for Astronomy 101 Courses, https://www.oercommons.org/authoring/18513-a-compilation-of-free-laboratory-activities-for-as/view.
- Moore, M. G. 2016, Teaching Science Online: Practical Guidance for Effective Instruction and Lab Work (Sterling, VA: Stylus Publishing).
- Online Astronomy Class from John Dutton at Penn State University with an example of how to facilitate a capstone project: https://www.e-education.psu.edu/astro801/node/2072.

ASS | IOP Astronomy

Astronomy Education, Volume 2
Best practices for online learning environments
Chris Impey and Matthew Wenger

Chapter 2

Who Are We Teaching Online, and Why?

Matthew Wenger

Distance learning is becoming a ubiquitous way of engaging with educational content, and online learning environments and their increasing capabilities are responsible for this growth in the last two decades. Higher education has begun to embrace online learning in the form of online classes as well as fully online degree programs. Online learning suits the needs of an increasingly diverse and mobile student population. It also gives students the flexibility to study without being tied to a campus and allows them to combine higher education with jobs and family life. In the developing world, online learning can be a democratizing influence that spurs economic development.

Learning Outcomes

After reading this chapter readers will be able to:
- Describe recent trends of online learners for college-level students.
- Give examples of how differences in motivations affect learning.
- Describe the different online learning environments.

2.1 Introduction

Even before health concerns forced instructors and students online in the spring of 2020, the use of online courses was controversial. Some have criticized online classes as an inferior form of education (Allen & Seaman 2012), and there are concerns that they provide an isolated learning experience (Gillette-Swan 2017; Johnson 2017). Still others fear online learning is a homogenized, western-dominated colonization of education, while at the same time, it is considered by many to be a promising means to increase access to education in developing countries (Trines 2018). Online classes are also the preference of the growing fraction of students who are not resident on campuses or who must juggle education with jobs and family obligations. Additionally, current trends in education require that almost every university explore online education options.

Although distance learning, in the form of correspondence courses, has been around for nearly two hundred years, digital technologies have changed the way that students and instructors interact, and have increased the accessibility of learning opportunities around the globe. These online learning environments are complex because they are often intersections of learners from diverse demographics who often have varying goals and motivations. In this volume we will be focusing on college-level online learning. These experiences may take the form of individual courses at universities and colleges, entire online programs provided by universities and colleges, and alternative online experiences that may be used for credit but also may be open to lifelong learner anywhere in the world to take.

Digital education is flourishing, and now increasing at a staggering rate. Among European institutions of higher learning, the vast majority report that they are seeing interest and increased demand for more flexible provision of degree and non-degree education (Gaebel et al. 2018). The number of massive open online classes (MOOCs), for example, has skyrocketed since they first appeared in the 2000s (Allen & Seaman 2014). The New York Times declared 2012 "the year of the MOOC"—an acronym that was, at the time, still an unfamiliar term. Since then, the number of MOOCs has increased by more than 683%: according to Class Central, a MOOC listings provider, there are now 13,500 courses on offer worldwide compared with only 1200 MOOCs in 2013. The total number of learners enrolled in MOOCs has shot up to 110 million from 10 million (Shah 2019). Most MOOCs are offered directly by private providers like Coursera or edX, but the number of universities offering MOOCs has also increased from 200 to over 900. MOOCs are now mainstream, and the number of available courses is growing daily. While MOOCs traditionally have been free and open to all, major MOOC providers like Coursera are rolling out a growing number of for-credit classes. By 2019, there were 820 credentials like mini-Masters and 50 full undergraduate degree programs. MOOCs have traditionally had very low completion rates, under 10%, a level that would be unacceptable in most for-credit college classes.

2.2 Worldwide Online Education

Providing access to education is a global challenge. In 2016, 1 in 5 children worldwide did not participate in any form of educational experience (UIS 2018). The youth population in Africa alone is expected to double to 830 million people by 2050 (Chatterjee 2015) and there are few resources dedicated to educating this population (Trines 2018). Additionally, students across the world from varying demographic backgrounds are currently working to access education online. It is in this context that online education is getting increased attention as a method for broadening access to education in a system with severe cost constraints. Gates & Gates (2015), for instance, believe that online learning will revolutionize education in the developing world and help close global literacy gaps and are using money from their foundation to support these programs.

There are many challenges that must be overcome for online higher education to fulfill these lofty promises. Low Internet penetration, low public esteem for online learning, and lack of online educational repositories in Arabic language are major obstacles to growth in Middle Eastern Countries (Palvia et al. 2018). Even with these persistent technological barriers, current trends in Sub-Saharan Africa and South Asia illustrate that online education is nonetheless gaining traction in these regions. The spread of smartphones means that digital learning is a more viable option for many people in developing countries, and mobile broadband technology is quickly penetrating even remote rural regions, providing Internet access to the people that live there (Bahaia & Suardi 2019). This growth is occurring not necessarily because online programs represent a better form of learning, but because it is seen as a cost-effective means to increase access to educational opportunities (Trines 2018). Similarly, countries like India are embracing online education due to population growth and exploding demand for education (Trines 2018; IGNOU 2020). Open distance education universities in Bangladesh, India, Iran, Pakistan, South Africa, and Turkey currently enroll more than 7 million students combined (Trines 2018). Ghana, South Africa, and Malawi lead the online education movement in the African continent (Palvia et al. 2018), and financially struggling governments in low-income countries see online education as a cheap option to quickly bridge capacity gaps. Whether or not online education can live up to this promise remains to be seen. However, the growth potential for online education in developing countries is certainly enormous (Trines 2018).

2.3 Western Countries

Most of what we know about online education comes from efforts in western countries. In the United States of America, enrollments in postsecondary education were growing overall until 2012 when they began to decline (Seaman et al. 2018). Although most institutions saw a moderate decline in on-campus enrollments, the largest declines came from for-profit institutions due to changes in federal rules that limited the use of student financial aid. Even as overall enrollments in postsecondary education dropped, participation in online classes grew by 5.7% between 2016 and 2017, according to the National Center for Education Statistics (Ginder et al. 2018). This represents the fourteenth year in a row that enrollment in distance learning has increased (Allen & Seaman 2017; Seaman et al. 2018) and exceeds the gains seen over the previous three years (Allen & Seaman 2017). According to the 2018 Science & Engineering Indicators from the National Science Board 14% of undergraduate students were enrolled in an online only program, and a further 15% were enrolled in both on-campus and online courses. The data also show that exclusive enrollment in online classes is considerably higher at for-profit colleges and at the graduate level. The National Center for Education Statistics (NCES) shows that in 2017, 49% of the 1.3 million undergraduate students attending private for-profit institutions were enrolled exclusively in distance education courses, while those numbers were 19% for students at private non-profit institutions and 11% for students at public institutions. Allen & Seaman (2017) report that 31.6% of students take at least one distance education course. In addition, 80% of administrators say that demand for online classes is increasing (BestColleges 2018).

Distance students are fairly evenly split between those who take both distance and non-distance courses (3.3 million students) and those who take exclusively distance courses (3.0 million; Allen & Seaman 2017). The number of students studying on a campus has dropped by over one million (1.2 million, or 6.4%) between 2012 and 2016 (Allen & Seaman 2017). Public institutions command the largest portion of distance education students, with 67.8% of all distance students (Allen & Seaman 2017). Distance education enrollments are highly concentrated, five percent of institutions account for almost half of all distance education students (Allen & Seaman 2017). Distance enrollments remain local: 52.8% of all students who took at least one distance course also took a course on-campus, and 56.1% of those who took only distance courses reside in the same state as the institution at which they are enrolled (Allen & Seaman 2017). Virtually no distance enrollments are international: only 0.7% of all distance students are located outside of the United States (Allen & Seaman 2017). Worldwide, it is likely that online learning will become a mainstream part of the educational infrastructure by 2025 (Palvia et al. 2018), with the 2020 health circumstances accelerating this transition.

2.4 Demographics and Motivations

The rise of online education is resulting in increased access to higher education. This means that the students who take online classes are often different from traditional undergraduate students. It is, however, difficult to generalize online students as a whole, because there is so much variation between students in different programs. Several studies have shown that fully online students tend to be employed at the same time they are taking online classes and they are often part-time students. Not all online learners, however, fit into this mold. Other research has found cohorts of rural, white, low-income women, 1/3 of whom received financial assistance (Johnson 2015). A BestColleges survey of online students and administrators found that roughly 25% of responding administrators reported their students trended "older," and 20% reported students trending "younger." Although these various reports may seem at odds with each other, Johnson (2015) observes that the differences in student demographics between studies is often due to variations between program offerings and researcher sampling techniques. These variations may be explained by understanding differences between student motivations for taking college courses, and the courses and programs at institutions that draw these different populations of students. Ladd et al. (2014) describe six categories of learner motivations, that also have a relationship to learner age:

College Student Characteristics and Motivations	
Category	Description
Aspiring Academics	18–24 years old, focused on academic studies
Coming of Age	18–24 years old, exploring college academics, social offerings, and a variety of activities

Academic Wanderers	Older students who perceive the advantages of a college degree, but are unsure about academic and career goals, and how to reach them
Career Starters	Wider age range, interested in college as a path to a specific career
Career Accelerators	Older students with some college and job experience, interested in college as a way to move forward in their current career field
Industry Switchers	Older students with some college and job experience, interested in transitioning to a new career field

Based on this model, it is important for online instructors to understand the goals of the programs in which they are teaching. Different students with different motivations and cultural backgrounds will choose different academic programs based on their particular goals. The BestColleges survey found that a large majority of the students surveyed (77%) were taking online classes for career-related reasons, with 37% identifying as career accelerators, interested in advancing in a career where they already work or have experience, and 37% were industry switchers, interested in changing careers and entering a new field.

The most recent BestColleges survey (2020) also found that students who responded to their survey were most commonly female (56%), between the ages of 24 and 44 (58%), taking undergraduate-level courses (57%), had at least one child (60%), Caucasian (66%), employed (56%), low-income (50%), and enrolled as full-time students (56%). A small percentage (12%) of administrators who responded said they are seeing increased diversity in their online students across gender, age, and ethnicity. Students reported that the convenience and flexibility of online programs is the main reason for enrolling in online courses, with 57% of students reporting that they had visited their school's physical campus, a decrease from 63% the previous year (BestColleges 2020).

The previous BestColleges survey (2019) noted that 16% of responding schools reported that they are seeing increases in the number of students who are enrolling from outside the USA, from out of state, or at greater distances from campus than previously seen. The majority of online students, however, were found to be "hyperlocal," in other words, within easy commuting distance of campus. Some of these students were taking classes on-campus as well, and others could attend classes on campus, but choose not to. Many of the resident students indicated they were taking online courses to supplement course schedules and reduce time to graduation. Reducing the time to degree completion is a recurring theme in those who choose to study online. This same 2019 report also showed that responding schools noted that online classes had seen an increase in students with disabilities, learners for whom English is a second language, underrepresented minorities, and economically disadvantaged students.

Although most student respondents to the BestColleges (2020) survey (31%) reported no concerns, 24% expressed concerns about the quality of instruction and academic support, and 21% expressed concerns about the perception of online degrees by prospective employers and 15% had concerns about the lack of community and/or interactions with professors and classmates.

These demographics and motivations are similar to those of the learners who enroll in courses that are not for credit, including MOOCs. A 2013 White paper about MOOC learners on the Coursera platform found that learners were young (under 40), well-educated (had bachelor's degrees or advanced degrees), and over half were male and from developed countries (Christensen et al. 2013). Research that the authors have conducted on an astronomy MOOC through Coursera showed that 50% of participants who completed an in-class survey were under the age of 40 and 37% were over the age of 40. The MOOC students in this astronomy class were also well-educated, with 69% of students having a bachelor's degree or higher. Although some of these learners (8%) were retired, the majority were working professionals (73%) (Impey et al. 2016). The astronomy MOOC students, however, were different from many Coursera students in that most were hobbyists or pursuing an interest in the subject of astronomy, rather than looking for professional development to further their careers (Formanek et al. 2019), demonstrating that it is insufficient to make sweeping assumptions about an online population, and that it is critical to find out more about students based on the particular institution, program, and classes being taught.

2.5 Conclusion

Online Education is steadily gaining ground in the higher education landscape, to the point where most students in the United States of America take one or more classes online each year. Its growth is fueled by the spread of broadband Internet and the maturation of platforms to distribute video lectures and other instructional materials to large numbers of students. The demand for online learning in the United States is driven by demographic changes in the population seeking higher education. These changes include a growing number of non-traditional students who are older or have jobs and families, and the need for retraining in an evolving job market.

The level of a person's education is tied to economic opportunity, and this is particularly relevant in developing countries. Globalization of online education can happen only if there are standard technology platforms (like the Internet), bridging of the digital divide, accommodation of diverse languages and cultures, standard curricula, uniform evaluation processes, and mechanisms for transferring academic credit between different countries (Palvia et al. 2018).

Online education is increasingly important and this volume talks about best practices in teaching in online environments. Astronomy is well-placed to take advantage of the capabilities of online learning since the richness of the subject can be conveyed clearly with digital media. However, the online environment poses particular challenges for maintaining student engagement and for supporting the needs of a diverse learner audience. Now, more than ever it is important to not only understand who is interested in online learning, their motivations, and challenges, but also understanding how to best serve these students.

References

Allen, I. E., & Seaman, J. 2012, Conflicted: Faculty and Online Education
Allen, I. E., & Seaman, J. 2014, Grade change: Tracking online education in the United States
Allen, I. E., & Seaman, J. 2017, Digital Compass Learning: Distance Education Enrollment Report 2017
Bahaia, K., & Suardi, S. 2019, The State of Mobile Internet Connectivity Report 2019, GSMA Intelligence, https://www.gsma.com/mobilefordevelopment/wp-content/uploads/2019/07/GSMA-State-of-Mobile-Internet-Connectivity-Report-2019.pdf
BestColleges 2018, 2018 Online Education Trends Report, https://res.cloudinary.com/highereducation/image/upload/v1/BestColleges.com/BestColleges.com-2018-Online-Trends-in-Education-Report.pdf
BestColleges 2019, 2019 Online Education Trends Report, https://res.cloudinary.com/highereducation/image/upload/v1556050834/BestColleges.com/edutrends/2019-Online-Trends-in-Education-Report-BestColleges.pdf
BestColleges 2020, 2020 Online Education Trends Report, https://res.cloudinary.com/highereducation/image/upload/v1584979511/BestColleges.com/edutrends/2020-Online-Trends-in-Education-Report-BestColleges.pdf
Chatterjee, S. 2015, Promise or Peril: Africa's 830 Million Young People by 2050, UNDP-United Nations Development Program, http://www.africa.undp.org/content/rba/en/home/blog/2017/8/12/Promise-Or-Peril-Africa-s-830-Million-Young-People-By-2050.html
Christensen, G., Steinmetz, A., Alcorn, B., et al. 2013, The MOOC Phenomenon: Who Takes Massive Open Online Courses and Why? http://dx.doi.org/10.2139/ssrn.2350964
Formanek, M., Buxner, S., Impey, C., & Wenger, M. 2019, PRPER, 15, 020140
Gaebel, M., Zhang, T., Bunescu, L., & Stoeber, H. 2018, Trends 2018: Learning and Teaching in the European Higher Education Area (Brussels: European University Association)
Gates, B., & Gates, M. 2015, Our Big Bet for the Future: 2015 Gates Annual Letter, https://www.appropriations.senate.gov/imo/media/doc/hearings/032615%20Gates%20Foundation%20Annual%20Letter%202015%20-%20SFOPS.pdf
Gillette-Swan, J. 2017, J. Learn. Des., 10, 20
Ginder, S. A., Kelly-Reid, J. E., & Mann, F. B. 2018, Enrollment and Employees in Postsecondary Institutions, Fall 2017; and Financial Statistics and Academic Libraries, Fiscal Year 2017: First Look (Provisional Data) (NCES 2019- 021rev) (Washington, DC: US Department of Education)
Impey, C., Wenger, M., Formanek, M., & Buxner, S. 2016, CAP, 21, 20
Indira Gandhi National Open University (IGNOU) 2020, Preamble, http://www.ignou.ac.in/ignou/aboutignou/profile/2
Johnson, A. 2017, Why Virtual Teaching Will Never Ever Replace Classroom Teaching, https://study.com/blog/why-virtual-teaching-will-never-ever-replace-classroom-teaching.html
Johnson, G. M. 2015, JUTLP, 12, 4
Ladd, H., Reynolds, S., & Selingo, J. 2014, The Differentiated University: Recognizing the Diverse Needs of Today's Students, The Parthenon Group, https://www.luminafoundation.org/files/resources/the-differentiated-university-wp-web-final.pdf
National Science Board 2018, Science and Engineering Indicators 2018, NSB-2018-1 (Alexandria, VA: National Science Foundation)
Palvia, S., Aeron, P., Gupta, P., & et al., 2018, J. Glob. Inf. Technol. Manag., 21, 233

Seaman, J. E., Allen, I. E., & Seaman, J. 2018, Grade increase: Tracking distance education in the United States, https://onlinelearningsurvey.com/reports/gradeincrease.pdf

Shah, D. 2019, By the numbers: MOOCs in 2019, Class Central, https://www.classcentral.com/report/mooc-stats-2019/

Trines, S. 2018, Educating the Masses: The Rise of Online Education in Sub-Saharan Africa and South Asia, https://wenr.wes.org/2018/08/educating-the-masses-the-rise-of-online-education

UNESCO Institute of Statistics (UIS) 2018, One in Five Children, Adolescents and Youth Is Out of School, http://uis.unesco.org/sites/default/files/documents/fs48-one-five-children-adolescents-youth-out-school-2018-en.pdf

Chapter 3

Effective Course Design

Adrienne J Gauthier

This workshop-in-a-box style chapter will guide instructors and education professionals through an integrated course design method. What will students know and be able to do with that knowledge by the end of the learning experience? How will you know that they are successful and help them monitor their progress along the way? What will students be doing, practicing, and experiencing while learning? You will answer these questions and more by exploring resources, brainstorming, writing reflections, and completing worksheets. This chapter has a tangible outcome—that you walk away with the instructional framework for at least one course goal. Find a colleague to go through the adventure with you, and you will have built-in feedback every step of the way. Let's get started!

Learning Objectives

By the end of this chapter, you will be able to
- discuss the benefits and challenges of using a structured course design method,
- brainstorm and describe factors that influence the learner and the learning environment,
- describe big-picture course goals and questions that will have enduring meaning and relevance for learners,
- articulate descriptive, realistic, and measurable learning objectives for your learners,
- discover and align assessment strategies and feedback loops between you, your learners, themselves, and the content,
- discover and align learning activities and experiences that engage the learner and help them successfully meet the learning goals and objectives, and
- communicate the flow and alignment of your course with learners and colleagues.

3.1 Introduction

Welcome to a task-oriented chapter on the basics of course design, a "workshop-in-a-box" experience that you can complete at your own pace. My hope is that you will walk away with a solid start on designing a new course or taking a fresh look at a course you have taught before. You will be asked to complete reflections and worksheets that mimic an introductory course design workshop series or an individual consultation between a professor and an instructional (or learning) designer.

You will not be asked to design your entire course in this chapter; instead, you will write one course goal and drill down into a unit or set of topics. Once you can see the entirety of the process through focusing on one course goal, you can then go back and work more effectively on your whole course. You will skim the surface of setting learning goals and objectives, designing assessments, creating feedback strategies, and developing learning activities. You will be provided with additional resources that include specific activities, ways to provide content to students, how to build effective assessments, and deeper information on how people learn. Take from this chapter what you will and use the ideas however they fit best for your course, students, and teaching philosophy.

Why devote the time and energy to a structured course design process? Summarized from the voices of past workshop participants, *the course design process has helped me:*

- have a clear vision of the course from the start,
- decide what to keep or remove from an overstuffed and content-heavy course,
- figure out where active learning would work best,
- find the flow between classes and topics,
- feel like everything in the course has intention and purpose, which made the learning better and deeper,
- gain clarity in the learning objectives and course goals, which made figuring out assessments easier, and
- communicate the rigor and robustness of the course to other professors, the department, and institution.

Additionally, many institutions require online faculty to review their courses against quality standards. Sometimes these are peer-reviewed by faculty colleagues using a tool like Quality Matters (www.qualitymatters.org/qa-resources/rubric-standards/higher-ed-rubric), and other times they are are formally assessed by an online teaching center. Online instructors can also self-assess during the design process by using tools like the Online Learning Consortium's free Quality Scorecard Suite (https://onlinelearningconsortium.org/consult/olc-quality-scorecard-suite/) or OPEN SUNY's OSCQR Course Design Review Scorecard (https://onlinelearning-consortium.org/consult/oscqr-course-design-review/). Most online course quality rubrics include sections on the alignment of learning objectives, activities, assessment, and feedback strategies. The course design process presented in this chapter will help set you up for success in meeting those guidelines.

Your students will also have some major benefits. When the course plan is explicitly shared with students on a course page or in your introductory course video, they not only know what you expect of them but have a vision for how to get there. They will be able to see the relevance in what you are asking them to do and how one assignment flows into the next and helps them learn.

Types of activities to work through:
- Reflect: Strategically placed reflection prompts help prep you for a topic by revealing what you already think, feel, and know. Use whatever you like to complete these prompts—pen and paper, typing into a document, or audio recording your responses.
- Task: You will be asked to complete the steps of course design in an organized and systematic way. The tasks are intended to be done in order as presented. Most tasks refer to a guiding worksheet, while others might include brainstorming and exploring external resources.
- Worksheets: Worksheets can help guide and organize your work. Use whatever method works best for you as you go through the steps. You can download all worksheets (.docx and .pdf) from the Course Design Workshop companion site: https://sites.google.com/view/cdw-companion.

Before you dig into this chapter, please consider the following:
- Working through the main steps in this chapter and focusing on one course goal could take you 10–15 hours, spread over a week or two. However, breezing through in more of a brainstorming and draft mode will take less time (4–6 hours) and give you a broad overview of the process.
- Find your teaching and learning center at your campus, or another group that might have instructional designers, learning designers, course designers, or faculty developers. Having someone to help guide you through the course design process or give feedback on what you are working on is invaluable.
- Recruit colleagues to join you in this adventure! They do not need to be in your department or your institution, but they should be working on a course. You can work through this chapter together and have frequent feedback from a peer.
- Timing! A structured course design method takes a lot of time, energy, and iteration. Please consider your available time before embarking on a full course design project and how far before the term you will need to start.

Reflect.

Please outline your own steps and process when you are creating a new course or getting ready to teach. For the reflections, use whatever you have available to you—computer, pen, and paper, etc. Be sure to save your notes as you will be asked to refer to them throughout the chapter.

3.2 What is Your Teaching and Learning Philosophy?

Please take a moment to think about your approach to teaching and learning. In this online Faculty Focus article,[1] Haave (2014) asks a few questions that will help you examine your teaching and learning philosophy. Do you know where you fall on the spectrum of teacher-centric to learner-centric environments? Later, you will be asked to answer Haave's questions.

This chapter reflects a learner-centered perspective, which means that it focuses on what you want the learners to be able to do, understand, and experience, not what you (the instructor) will be doing or teaching. You will be asked to consider what your students will need in order to meet the learning goals of the course, and then design your course with those needs in mind. Weimer (2012) outlines how an instructor might embody learner-centered teaching strategies. Please read this Teaching Professor blog post[2] to get more in-depth examples and explanations.

Weimer (2012) describes the effective teaching strategies of learner-centered teaching as follows:
1. ... engages students in the hard, messy work of learning,
2. ... includes explicit skill instruction,
3. ... encourages students to reflect on what they are learning and how they are learning it,
4. ... motivates students by giving them some control over learning processes, and
5. ... encourages collaboration.

Reflect.

What are some learner-centered things you already do in your teaching? What are you asking the students to do to help them learn? How are they interacting with the course content and each other?

Task.

Investigate your teaching philosophy by completing *Worksheet 1: Teaching Philosophy*. Download the worksheets from here: https://sites.google.com/view/cdw-companion.

---

[1] https://www.facultyfocus.com/articles/philosophy-of-teaching/six-questions-will-bring-teaching-philosophy-focus/.
[2] https://www.teachingprofessor.com/topics/teaching-strategies/active-learning/five-characteristics-of-learner-centered-teaching/.

3.3 Course Design Overview

The course design method in this "workshop" can be guided by these questions:
- What will the learner know and be able to do with that knowledge by the end of the learning experience?
- How will they (and I) know what they are able to do (and not yet do)?
- What will they be doing, practicing, and experiencing while learning?
- How can I communicate my course plan to learners in an approachable way?

You will aim to answer these questions while participating in the activities in this chapter. Learning design consultations with instructors can take many avenues and are full of questions and reflections to help the instructor gain clarity and think from alternate viewpoints. Obviously, this chapter cannot replicate a one-on-one consultation; however, there are multiple opportunities presented for you to stop, take a step back, and be reflective on what you are creating. If you can find a colleague to partner with on the work in this chapter, it would be of great value to discuss each other's course design plans.

3.3.1 Course Design Methods

This chapter relies on two popular course design models: *integrated course design* (Fink 2013) and *backward design* or Understanding by Design (Wiggins & McTighe 2008). Backward design was created for the K–12 world, is somewhat prescriptive, but has a lot to offer higher education. Integrated course design is less prescriptive, but still relies on basic principles and philosophies. A common thread between them is the focus on the learner. Figure 3.1 presents the primary principles of each method and outlines the steps in this chapter.

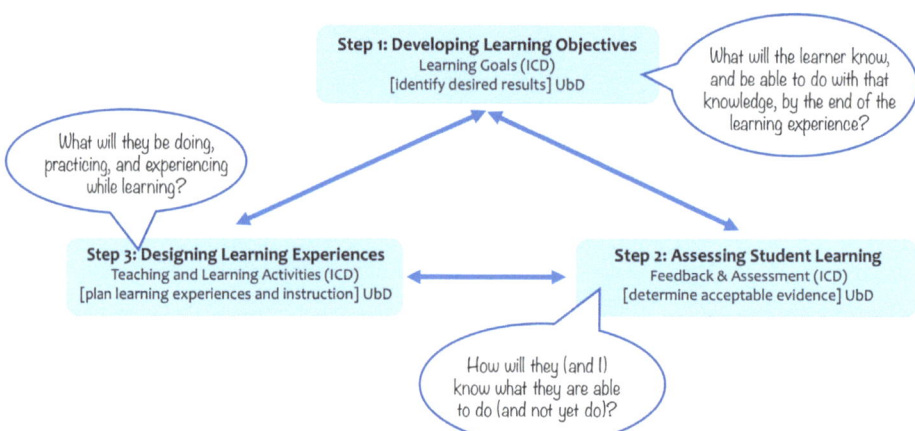

Figure 3.1. A merger of integrated course design and backward-design methods. Integrated Course Design (ICD; Fink 2013) and Understanding by Design (UdB; Wiggins & McTighe 2008) are very similar and will be the sequence followed in this chapter.

The starting point in Figure 3.1 is "Developing Learning Objectives." Once you are able to articulate *what the learner will be able to do*, the process pivots to "Assessing Student Learning." It might not be intuitive to think about how you might assess or evaluate students before you have come up with the learning activities. Following Fink's (2013, p. 71) advice, "creating the assessment activities first, greatly clarifies and facilitates answers to the question of what the learning activities need to be". After defining learning objectives and identifying assessment and feedback strategies, you will be asked to design activities and experiences that will help the learner practice skills, connect concepts, and think deeper about the course topics. When the assessments directly evaluate if the student has met the learning objectives, and the activities are relevant to practicing the skills and knowledge they will be assessed on, then we would say the course is *aligned*. You will work on recognizing when this is happening, how to make it happen, and how to communicate it to students.

The course design process is iterative and frequently loops back to prior elements. There can be a lot of back and forth thinking about the different aspects of the learning experience for the students, which can mean revisiting learning objectives and assessments after you think they are completely designed. You might also be thinking about activities when brainstorming assessments, and as you write learning objectives, you will have ideas for both activities and assessments. When focusing on one area of the course design method, you can jot down any extraneous ideas on a sticky note for use later.

Presented below is an example from an introductory astronomy course using the course design method in this chapter. This example focuses on only one unit related to one of the course-level objectives and only shows a snippet of the learning objectives, assessments, and activities. It is not a complete unit of a course, and there are concepts and activities not shown in this brief example. There are also areas that could be improved—what catches your eye?

Example
Course: Astronomy 1 "The Universe"

Related Course-level Learning Objective: Elaborate on how astronomers "see" and study astronomical objects through multi-wavelength light, data, and theoretical models.

Unit-level Learning Objective: Create visual and text materials that will communicate information and relationships about multi-wavelength light, astronomical objects, and how we "see" the universe, suitable for a general public audience.

Topic-level Learning Objectives:
Acquiring Knowledge (low cognitive level)
- Identify the different categories of light, from low energy to high energy.
- Recall the relationships between the temperature, wavelength, energy, and frequency of electromagnetic radiation.

Assessments	Activities	Feedback
(Formative) Answer end-of-chapter questions in the online textbook website. Just in time scoring and hints.	Watch animations and videos on textbook website to learn basic concepts	Textbook website auto-graded and gives immediate feedback and hints.
(Formative) Low-stakes "knowledge check" online quizzes	Participate in think-pair-share style activities during a live class session in Zoom	Discussion with peers and through polling results
(Formative) Submit 'muddiest point' responses.	Reflect at the end of the module	Instructor feedback at the beginning of next live session or in online video recap

Making Meaning (mid cognitive level)
- Interpret color scales in multi-wavelength imagery (Earth and astronomical).
- Compare and contrast multi-wavelength images of the Sun by relating temperature, energy, and wavelength of light in the imagery.

Assessments	Activities	Feedback
(Formative) Low stakes knowledge check quizzes	Participate in think-pair-share style activities during a live class session in Zoom Infrared Zoo/Everyday Objects lesson	Discussion with peers
(Summative) Group summary submissions in Canvas	Asynchronous discussions in small groups, worksheets/tasks	Discussion with peers in small groups Comments on group worksheets in Canvas

Transfer of Knowledge (high cognitive level)
- Analyze multi-wavelength images of the same astronomical object and relate detection method/telescope to the observation.

Assessments	Activities	Feedback
(Formative) Submit progress report and outline.	Team contracts/reflection points Synchronous and asynchronous discussions in small groups	Feedback on team outline in Canvas Feedback on team dynamics
(Summative) Create a multimedia blog/news post. Graded (rubric) and final peer group feedback. Submit individual reflection.	Research objects/telescopes, work in small teams, develop narrative, create blog post	Peer feedback given back to teams. Graded via rubric with comments.

3.3.2 Course Design Prep

Before moving forward into the course design steps, please consider the Situational Factors[3] (Fink 2005a) related to your course and students. Situational factors are things like logistical details, institutional expectations, real-world relevance, nature of the content discipline, what the learners are like, and what your own teaching experiences have been (Fink 2005a). Answering these questions will be your lighthouse to guide you through designing a relevant and robust course for you, your learners, and your institution.

> *Task.*
>
> Complete *Worksheet 2: Situational Factors*[4] and prep your frame of mind by thinking about qualities of the learners, learning environment, institution, and your own teaching experiences.

> *Reflect.*
>
> In the Introduction, you were asked to reflect on how you usually design a course or get ready for a term of teaching. In reviewing the course design method in Figure 3.1—in what ways is it similar to your process? What challenges do you think you might have trying out this model?

---

[3] https://www.ideaedu.org/Portals/0/Uploads/Documents/IDEA%20Papers/IDEA%20Papers/Idea_Paper_42.pdf.
[4] https://sites.google.com/view/cdw-companion/worksheets.

3.4 Step 1: Developing Learning Objectives

In Step 1: Developing Learning Objectives, you will brainstorm at least one overarching course goal and express the big ideas or "enduring understandings" of your course. You will write clear course-level, unit-level, and topic-level learning objectives, and sort them into categories that describe depth and rigor. This is usually the longest step in the process as this is truly the foundation of everything else in your course. It is worth the effort.

3.4.1 Big Ideas, Goals, and Questions

Have you thought about the big ideas and overarching questions of your course? You might have these written in the course description or course goals section of your syllabus. If it is a new course you are working on, take this time to brainstorm and figure out what you want the learners to ultimately take away from the experience. Wiggins & McTighe (2008) refer to these ultimate takeaways as enduring understandings. What do you want students to be able to remember a few years later? What are the broad understandings that connect pieces of course content to each other and to their lives beyond your class? It is easy to get caught up in the microtopics of a course, when the true nature of the course might have a broader context. Do you care if students remember how to classify galaxies, or do you instead want them to have a good understanding of the process of scientific discovery throughout history?

Another way to think about and phrase your course goals might be as "essential questions." Wiggins & McTighe (2013) characterize an essential question as being "open-ended…thought-provoking…intellectually engaging…higher-order thinking…important, transferable ideas…raises additional questions…requires support and justification…and recurs over time."

When creating course goals and big ideas in a learner-centered way, they are less about what you will teach students, and more about what students will be able to do, describe, answer, and understand for longer than the end of the term. Focus on what you want your students to understand about the content, at a deep level, and how they might connect ideas together.

If your course is part of a sequence for majors in your department, find out what the goals and objectives are for the other classes to see where your course fits into the student experience. If your course is meant to fulfill a general interest or science distributive at your institution, your long-lasting goals might not be the actual content at all. Your overarching course goal might be more about the nature of science or science and society. How does your course fit into the overall student experience at your institution?

You will be reminded throughout this chapter that course design is a highly iterative process. You might come back and change your course goals and essential questions a little (or a lot!) as you get deeper into the details of units or activities later in the chapter. You might realize that what you thought was a course goal is not really addressed. Or, you might discover other important understandings that deserve to be a course goal. Be prepared and open to discovering missed opportunities or the need to remove course content.

Example course goals and big-picture questions:
- How does the scientific understanding of our universe change over time?
- Understand different scales in the universe, from micro to macro, from DNA to large-scale structure.
- How do scientists "see" and observe the universe?
- Appreciate the different types of astronomical objects and how they show an evolving universe as astronomers look back in time.
- What does it mean when we say "telescopes are time machines"?
- How are scientists creative in their work?

Course Goals Guide.

Consider these questions as you write and review your course goals, enduring understandings, and/or essential questions. Use whichever format feels right to you.
- *Are they written in nonexpert jargon?*
    Course goals, big ideas, and central questions should be approachable to nonexperts. Survey courses might use different language from an advanced course. Take into account the background knowledge students will have coming into your course.
- *Do they describe connections between course concepts or connections to real-world (outside of class) contexts?*
    Students should clearly see how your course connects to the world around them and how it might affect their own personal worldview or life. Why should they care or be interested in your course?
- *Are they lofty or vague?*
    Sometimes, the "big picture" can come off as vague. Students should be able to read your course goals and, without too much effort, know what they will experience and get from your course.
- *Do they inspire curiosity and promote an intellectually engaging learning environment?*
    Get them curious by using active and descriptive language.

Task.

Use *Worksheet 3: Course Goals*[4] to think through and write out your goals and/or big-picture ideas, which will in turn focus how you work on the course design steps in this chapter.

3.4.2 Significant Learning

Now that you have some course goals in mind, take a moment to reflect on Fink's Taxonomy of Significant Learning (2013, 2005b). Fink (2013, p. 7) presents a way to think about the different types of learning that can happen in a course beyond the course content itself. Fink summarizes the teaching philosophy as "teaching should result in something others can look at and say, 'that learning experience resulted in something that is truly significant in terms of the students' lives'".

Are your course goals only about ways for students to understand and apply course content, or do your goals encourage students to make meaningful connections to the world around them? Are learners going to be highly engaged and think critically about how the content is personally relevant? Take a look at descriptions of the types of significant learning and then revisit your course goals, enduring understandings, or essential questions. Do any fall into Fink's taxonomy, or can they be modified to capture the essence of "significant learning"? If you are not ready to dive into Fink's philosophy yet, just move onto the next section on learning objectives. You might get some ideas as you move through the course design steps.

Fink's Taxonomy of Significant Learning
Fink (2005a, p. 2) states, "One important feature of this taxonomy is that each kind of learning is interactive. That is, each is able to stimulate any of the other kinds of learning. For example, "Foundational Knowledge" may stimulate "Critical Thinking," which in turn may stimulate "Connecting Ideas," encouraging one to "Learn About Oneself," etc. The intersection of these interrelated kinds of learning defines "Significant Learning," the purpose of the Integrated Design process". Fink (2005a, p. 3) describes his Taxonomy of Significant Learning as the following:

- *Foundational Knowledge.*
    What key information (facts, terms, formulae, concepts, principles, relationships, etc.) is/are important for students to understand and remember? What key ideas or perspectives are important in this course?
- *Application.*
    What kinds of thinking (critical, creative, practical) are important for students to learn? What skills are required? Should students be expected to learn how to manage complex projects?
- *Integration.*
    What connections should students recognize and make among ideas within this course? Among information, ideas, and perspectives from this course and those in other courses or areas? Between material in this course and the students' personal, social, and/or work life?
- *Human Dimension.*
    What should students learn about themselves? What should they learn about understanding others and/or interacting with others?
- *Caring.*
    What changes/values should students adopt? Should interests be affected? Feelings? Commitments?

- *Learning How to Learn.*
    What should students learn about how to be good students in a course like this? How to learn about this specific subject? How to become a self-directed learner (developing a learning agenda and a plan for meeting it)?

3.4.3 Tutorial: Writing Learning Objectives

The language of learning objectives (LOs) inherently promotes a learner-centered philosophy. Earlier, this was described as thinking about what the learner will be able to do, which is different from what an instructor will teach. You will work on a hierarchy of LOs for your course, starting with course-level LOs. Organized below your course-level LOs will be unit-level LOs, and under those will be topic-level LOs. It might be that your course is better organized as a concept map, where multiple unit-level LOs feed into a bigger course-level LO. For courses that have a central theme, such as "the nature of science," students might visit that topic throughout the course, and various levels of LOs might look more like a tapestry.

There are a variety of tools and methods for crafting LOs so that they are clear, explicit, and easy to communicate to others. Sharing your course goals and LOs with students will make your teaching more transparent and can promote increased engagement and ownership of learning. The Eberly Center for Teaching Excellence[5] summarizes the benefits of sharing LOs with students perfectly:

> "To become self-directed learners, students must learn to (a) assess the demands of the task (in other words, the learning objectives/outcomes), (b) evaluate their own knowledge and skills, (c) plan their approach, (d) monitor their progress (which they can only do if they understand the type of knowledge they are expected to gain), and (e) adjust their strategies as needed."[6]

In this tutorial, you will use a popular cognitive learning taxonomy named Bloom's Revised Taxonomy (Anderson et al. 2001) as your main tool for articulating the levels of learning in your course. In the future, you might explore other tools or methods that are more aligned with your philosophy and course content. For example, there is a robust taxonomy that aligns with Fink's Significant Learning (2013) in Barkley & Major (2016).

Bloom's Revised Taxonomy
What is meant by learning taxonomy? A learning taxonomy is a way of describing the complexity, depth, or level of thinking students will encounter. The categories of the taxonomy can help you gain clarity in what you are asking students to do, and if it is at the right level. Learning taxonomies are not all theory, but come with

---

[5] The Educational Value of Course-level Learning Objectives/Outcomes, https://www.cmu.edu/teaching/resources/Teaching/CourseDesign/Objectives/CourseLearningObjectivesValue.pdf.
[6] See the full resource here: https://www.cmu.edu/teaching/resources/Teaching/CourseDesign/Objectives/CourseLearningObjectivesValue.pdf.

resources like action verbs to help the learning designer categorize the cognitive processes of the learner.

In Bloom's Revised Taxonomy (Anderson et al. 2001), there are six categories or levels of cognitive process. When using categorized action verbs to assist with course design, it is easy to get caught up in what cognitive level a specific verb belongs in. Though the action verb "identify" is listed in the lowest cognitive level of Bloom's in many resources, it might be that asking students to "identify" something is a higher-level task that brings together many different concepts. For example: identifying wavelengths on a typical electromagnetic spectrum chart is a "memorizing" task and fairly low level. However, identifying exoplanets in a visualization of star brightness data might require advanced understanding of variable stars, detection methods, and limitations of observations. In that case, maybe there is a better and more descriptive action verb for finding exoplanets in visualized data. How does analyze or discriminate work better in that context?

The best way to understand how Bloom's taxonomy works is to start using it, which you will do in the next section.

Level of complexity	Category	Description (Anderson et al. 2001, pp. 67–68)	Example action verbs
Lower levels of thinking and doing	Remembering	Retrieving, recognizing, and recalling relevant knowledge from long-term memory.	Identify Define List
	Understanding	Constructing meaning from oral, written, and graphic messages through interpreting...classifying, summarizing...comparing, and explaining.	Describe Explain Relate
Intermediate levels of thinking and doing	Applying	Carrying out or using a procedure through executing or implementing.	Compare Manipulate Determine
	Analyzing	Breaking material into constituent parts, determining how the parts relate to one another and to an overall structure or purpose.	Distinguish Correlate Classify
Higher levels of thinking and doing	Evaluating	Making judgements based on criteria and standards through checking and critiquing.	Predict Infer Criticize
	Creating	Putting elements together to form a coherent or functional whole; reorganizing elements into a new pattern or structure through generating, planning, or producing.	Create Hypothesize Adapt

You can find examples of Bloom's and Fink categorized verb lists in these resources:
- Vanderbilt University's Center for Teaching: https://cft.vanderbilt.edu/guides-sub-pages/blooms-taxonomy/.
- Azusa Pacific University has a posted copy of a popular verb list, original source unknown: https://www.apu.edu/live_data/files/333/blooms_taxonomy_action_verbs.pdf.
- Northeastern University's Center for Advancement in Teaching and Learning Through Research has posted an extensive verb list: http://www.northeastern.edu/nuolirc/wp-content/uploads/2018/01/Blooms-Taxonomy-Handout.pdf.
- In Barkley & Major's (2016, p. 19) *Learning Assessment Techniques* book, there is a robust verb list organized into Fink's Significant Learning Taxonomy categories.

Transform Goals into Course-level Learning Objectives
Review your course goals from *Worksheet 3: Course Goals*[4] and turn them into course-level LOs using action verbs. Course-level LOs articulate what students will be able to do at the end of your course to demonstrate what they have learned and understood. Find inspiration from your course goals and/or essential questions as you rewrite using the LO language. Course-level LOs should be the highest level of challenge where students are synthesizing, creating, and transferring application of the understandings. Also consider noncognitive domain course objectives. These are things like reflecting on worldviews or the nature of science, communicating science to different audiences, and learning how to work in teams are all wonderful noncognitive objectives and are important in creating educated global citizens.

Examples of developing course-level LOs:

Enduring understanding/essential question	Potential course-level learning objective *By the end of the course, the student will be able to*
Understand different scales in the universe, from micro to macro, from DNA to large-scale structure	Construct and communicate a mental model of various scales of the universe and how they relate to each other.
How do scientists "see" and observe the universe?	Elaborate on how astronomers "see" and study astronomical objects through multi-wavelength light, data, and theoretical models.
What does it mean when we say, "telescopes are time machines"?	Construct a concept map of the universe, organizing astronomical objects by distance and evolutionary stages.
How are scientists creative in their work?	Discuss the importance of scientific inquiry in the context of real-world examples, what it means to be a skeptic, and how it impacts daily life.

As you write your course-level LOs, you will get ideas for how students will demonstrate their understandings to you, what they might produce or hand in, and what they might do as learning activities. Do not skip ahead; instead, use sticky notes to jot down your ideas and then set them aside for later.

You might be thinking, "so what do I do with my course goals and essential questions from Worksheet 3 now that I've rewritten them?" Keep them! Use them as your course description in your syllabus or online course site in the learning management system (Canvas, Blackboard, D2L, etc.) Your course goals are the heart of your course and will complement your newly written course-level LOs in a more narrative form.

Learning Objectives Guide.

Consider the following questions as you develop and review LOs (course level, unit level, and topic level):

1. *Do your LOs convey something about a single skill or understanding about the content?*
    Multiple skills or layers should be broken out into separate LOs. They should be able to answer *what*, *why*, and/or *how*.
2. *Do your LOs speak to an action or task?*
    LOs should use action verb language like *compare* and *contrast* or *analyze*. LOs should start with "the learner will be able to" which then implies that a verb or two will follow.
3. *Do your LOs describe specifically what the learner will be able to demonstrate?*
    The evidence of learning should be clear to anyone who reads the objective. This is what the instructor (and the learner) can assess and evaluate to know if the LO is met. Verbs like "understand," "appreciate," and "demonstrate" are vague and do not actually describe what the learner would do.
4. *Are they written in jargon-free language?*
    As novices in the subject matter, learners appreciate having approachable and understandable LOs. Introductory courses and senior-level majors' courses have different types of "novices," and so the language used would be different. Think about the audience when writing LOs. Generally, the more granular the LO (topic level versus course level), the more jargon you can use.

Task.

Write and review your course-level LOs using *Worksheet 4: Course-level Learning Objectives*.[4]

Unit-level Learning Objectives

The next step is to develop unit-level LOs. Unit-level LOs organize underneath the course-level LOs and are fairly high level. In some courses, topics might be broken out into themes which weave through different sections of the course over the term. If that is the case, pick a theme to work on for the rest of this chapter. As you gain clarity in the unit-level LOs, you might find yourself wanting to change the course-level LOs. Please do! This is part of the iterative course design process.

Examples:
By the end of the unit (or theme) students will be able to
- debate and defend the scientific, political, and environmental considerations for various observatories and telescopes;
- compose an explanation that will communicate what is meant by the phrase "the nature of science," that can interest, inform, and be relevant to a nonscientist/expert audience; and
- evaluate different exoplanet detection methods and relate to an online exoplanet citizen science experience.

Task.

Use *Worksheet 5: Unit-level Learning Objectives*[4] to outline, write, and review your unit-level LOs.

Topic-level Learning Objectives
Within each unit of your course, you might have a module or topic structure—something that lasts a few days or a week. You can even continue into more granularity by thinking about the activity or assignment level. Only go as far as you are engaged and interested to go in this "introduction to course design" experience. Even if you only write course and unit-level LOs, you are doing great and it will be a significant benefit to you and your learners.

For topic-level LOs, you will want to consider all levels of Bloom's Revised Taxonomy. There are times when learners need to first acquire the basics and be able to remember and recall content and facts. Learners will then apply those understandings to activities like solving a simple problem, where they will make meaning of what they have already learned. Sometimes, this intermediate level is as far as the complexity goes. Other times, learners will need to transfer their intermediate understanding and skill to new and complex situations in order to demonstrate mastery.

Wiggins & McTighe (2008) refer to this concept as AMT or acquiring knowledge (A), making meaning (M), and transfer (T). "A" is the lowest level, and "T" is the highest. It can be challenging to distinguish between making meaning and transfer. Think of transfer as bringing multiple ideas together in a new way or applying something in a novel context or situation.

An example of categorizing topic-level LOs as A, M, or T is a flipped course where students watch short videos prior to class in order to solve or discuss beginner- to intermediate-level problems during class. The topic-level LOs in (A) *acquiring knowledge* focus on basic knowledge that students learn from the video clips. Tasks like recalling a definition, duplicating a diagram, or matching answers to pictures fall into this category. During class, they would use the knowledge gained from the first exposure (video) to solve problems, answer conceptual questions, or discuss concepts with peers. Those would be (M) *making meaning* experiences. (M) ranges from entry-level conceptual questions through mid-level problem solving or analyzing data. (T) *Transferring* knowledge indicates the highest level of learning and usually happens near the end of a unit or after multiple concepts are learned. (T) is about synthesizing everything students have learned and directing that knowledge and understanding to a new context or project. In a fully online course, students might encounter the (M) and (T) objectives through small group online discussions, either asynchronous in a forum or synchronously during a live video session. An example of authentic transfer is to ask students in an upper-level astronomy course to design an actual observing plan for a research-grade telescope.

Examples of A, M, and T for one of our example unit-level LOs:

Unit-level LO: Evaluate different exoplanet detection methods and relate them to an online exoplanet citizen science experience.

A, M, T level	LO
(A) Acquiring knowledge	Describe different types of extrasolar planets.
(A) Acquiring knowledge	Match the type of extrasolar planet observation to a telescope facility.
(M) Making meaning	Compare and contrast the methods that scientists use to search for and observe extrasolar planets.
(M) Making meaning	Relate the habitable zone of an exoplanet system to our own solar system.
(T) Transfer	Distinguish between an observation of an extrasolar planet and various variable stars while presenting reasons for the findings.

Task.

Use *Worksheet 6: Topic-level Learning Objectives*[4] to outline, write, and review topic-level LOs. As you review, categorize each LO as A, M, or T.

Congratulations! You have completed Step 1!
Take a break from this chapter. Let Step 1 slosh around a bit in your brain. Review the worksheets, make any changes, or include additional ideas, and then continue on with Step 2 in a few days.

3.5 Step 2: Assessing Student Learning

In Step 2: Assessing Student Learning, you will review different types of assessment, think about the importance of feedback, research and find assessment methods, and check that the assessment strategies align with your LOs. Designing opportunities for assessment and feedback will help you and your students determine their progress toward, and success in meeting, your LOs.

> *Reflect.*
>
> Brainstorm a list of answers to this question, "what are assessments good for and/or used for?" Challenge yourself to list at least five things from the teacher perspective and then five things from the student perspective. Let none of your items be "for a grade."

3.5.1 Types of Assessment

Assessment of student learning should give both the instructor and the learner useful information on the progress toward meeting the LOs. Two types of assessment are *summative assessment* and *formative assessment*. In a summative assessment task, students are asked to demonstrate mastery of the LOs, usually at the end of a unit or as the final product of the course. Instructors will evaluate the level to which the student succeeds and might assign a grade or performance score. Common summative assessments are final exams, term projects, or the final submission of a paper. More innovative summative assessments could be creating multimedia projects, fabricating something that can help teach others, or reporting on a level-appropriate research project. It is the summary of what the student learned and experienced, and what they can show you they can now do.

The purpose of formative assessment is to monitor the progress of learning and give an opportunity for feedback. A classic style of formative assessment is referred to as classroom assessment techniques (CATs), a term and method coined by Angelo & Cross (1993). They describe what formative assessments accomplish:

> Their aim is to provide faculty with information on what, how much, and how well students are learning, in order to help them better prepare to succeed—both on the subsequent graded evaluations and in the world beyond the classroom. (p. 5)

When instructors design formative assessments, like CATs, into their courses, they "become better able understand and promote learning, and increase their ability to help the students themselves become more effective, self-assessing, self-directed learners" (p. 4).

The qualities of a CAT are also the good characteristics of any formative assessment strategy. A good formative assessment is "learner-centered, teacher-directed, mutually beneficial, formative, context-specific, ongoing, and firmly rooted in good teaching practice" (p. 4). Most CATs and other formative assessment techniques that were developed for face-to-face courses can be creatively reshaped to work in fully online courses.

Some popular and quick CATs are:

One-sentence summary (CAT 13; p. 183)
Students are prompted to write a single sentence that summarizes a topic or concept. It sometimes takes the form of *who–what–when–where–why–how*.

Example prompt:
In a single sentence, summarize one method that astronomers use to discover planets around stars (other than our Sun).

Invented dialogues (CAT 17; p. 203)
Students synthesize high-level course concepts into a creative narrative, which takes the form of a conversation or journal entry of a character.

Example prompt:
With a partner, write a series of short journal entries from the point of view of a cartoon character "photon" that is produced by a distant object and travels to Earth. Choose the wavelength of your photon character (infrared, visible, or X-ray) and briefly describe its origin, journey, and destination.

Misconception/preconception check (CAT 3; p.132)
Students are asked a series of questions that focus on common misconceptions for the content.

Examples:
Concept inventories (Bailey 2019) exist for many topics in astronomy—light, gravity, seasons, Moon phases, physics concepts, etc. Some teaching tools, like lecture-tutorials (Wallace & Prather 2019) and peer instruction (Mazur 1997) include activities to reveal misconceptions as well as help students overcome them.

Muddiest point (CAT 7; p. 154)
Students write a lingering question or point of confusion on a notecard or in an online survey. Twists on this include engagement questions or general feedback about the course or activities.

Example prompts:
- What are you still confused about on this topic?
- What was something new you learned in this activity?
- What would you be most excited to learn about in the next live video session?

Formative and summative assessments work in partnership throughout a course. If you find yourself not able to decide what the label is (formative or summative), then do not label it, but do articulate the purpose(s) it serves. For example, a quiz can be summative and formative. In a summative sense, the quiz rounds out the unit by bringing together the concepts at a high level to meet a (T) unit-level LO. However, it is only one milestone among many leading up to a final high-level and rigorous take-home exam. In that sense, the quiz is formative and just one step toward a higher course-level LO. You can gauge progress toward the end goal and students can get feedback on how to improve or confirm they are doing all the right things.

When designing assessments, focus on the following:
- What is the assessment is telling you, and the student, about their progress?
- What are you, and the student, going to do with that information?
- What type and format of feedback are you going to give?
- What is the learner expected to do with the feedback?
- How does the assessment fit into the bigger picture of the course-level LOs?

3.5.2 Feedback Loops

Assessments give you and the learners an opportunity to monitor progress and incorporate feedback. There are a variety of ways to incorporate feedback on what they have done well, where they are still struggling, and how to move forward in their learning to improve. Feedback might come from you, fellow students, teaching assistants, an online system, or self-assessment. Forward-looking feedback helps students figure out how they might improve or what their next achievable step might be. A benefit for using formative assessments and a robust feedback strategy is that students are prone to be more engaged and motivated in their learning (Nicol & Macfarlane-Dick 2006). You are setting up a learning environment that shows that you care about their learning and are consistently asking them how they are doing in their progress. It is this instructor presence and sense of community that helps to motivate students, particularly in fully online courses.

A powerful way to help students learn is to provide guidance on how to be aware of one's own learning. Thinking about what one knows or thinks, and using that to be more self-aware about one's progress in learning, is referred to as metacognition.[7] Examples of this are the following:
- At the end of a module ask students to write down what they are still confused about or what questions remain (Angelo & Cross 1993).
- Give an assignment where students reflect on their study strategies and their performance on an exam. They can also reflect on what they got incorrect and reasons why. https://www.cmu.edu/teaching/designteach/teach/examwrappers/.
- After an in-class polling question, task students to compare answers with a classmate and give reasons why they think their answer is correct (Carl Wieman Science Education Initiative (CWSEI) & The Science Education Initiative at the University of Colorado (CU-SEI) 2017).
- Promote a problem-solving strategy that requires students to first answer, "What is the problem asking?" and then write down "Here's what I know" and "Here's what I need to know" before putting pencil to paper to solve the problem (Angelo & Cross 1993).

---

[7] Metacognition, https://cft.vanderbilt.edu/guides-sub-pages/metacognition/.

When you examine learner progress regularly and frequently, you are better able to adjust instruction and give students what they need to move forward. You can also gain insight on where students get stuck or how an activity needs to change for next time. When students are asked to examine and reflect on their own progress and learning, they can have a clearer picture of what they do and do not understand, and where they need to focus their energy or get help.

3.5.3 Assessment Strategies

Using clear LOs as a guide, it is the instructor's responsibility to design an assessment that is relevant, meaningful, and will give feedback to the learner on their progress. For lower-level LOs in the (A) category discussed in Step 1: Topic-level Learning Objectives, a common formative assessment in an online course is to provide a low-stakes knowledge check quiz before a student can move forward in a sequenced module. There is only minimal feedback, and it is usually delivered through the quizzing tool as general question/answer feedback or a mini-tutorial. For higher-level LOs in the (M) and (T) categories, the assessments should be more rigorous and offer richer and personalized feedback. They should aim to be more authentic to real-world (outside of the course) contexts. It is worth noting that not all assessments need to be "graded" or even be reviewed by the instructor. Some assessment strategies have students self-assess against a rubric, discuss work with peers in the class, or use adaptive technology.

So where do you find assessment strategies you can use in your course? There are few go-to books and online resources in an educational developer's bag of tricks, and now we will build up your box of pedagogical goodies. Many of the resources below are written for face-to-face courses, and it is an opportunity to think outside the box and design assessments for the online environment.

Assessment Resources:
- *Classroom Assessment Techniques: A Handbook for College Teachers* (2nd ed., Angelo & Cross 1993). This book is rich in formative assessment activity implementation guides. The authors share their scholarly approach as background and justification for the 50 CATs techniques they outline.
- *Learning Assessment Techniques: A Handbook for College Faculty* (Barkley & Major 2016). In the same style as the CATs book, these authors provide robust context for 50 LAT techniques and guides.
- Carnegie Mellon University's Eberly Center for Teacher Excellence and Innovation provides overviews, justifications, and strategies for assessment in the college classroom, https://www.cmu.edu/teaching/assessment/index.html.
- UCF's Teaching Online Pedagogical Repository is rich with assessment ideas for the online classroom: https://topr.online.ucf.edu/assessment/.
- Internet search: "formative assessment techniques" or "formative assessment strategies" or "classroom assessment techniques." There are a plethora of resources online, and you might find more K–12 resources than college level. Do not be turned off by the K–12 flavor—much of what we do in higher

education filtered up from the K–12 educators. Learning is learning, no matter the grade level.
- PhysPort[8] supports various astronomy concept inventories and diagnostic tests. The PhysPort website can also help analyze your data with easy to use tools: https://www.physport.org/assessments/?Subjects=91&.

If you have a teaching and learning center at your institution, they will likely have the books above to borrow. Your library or interlibrary loan might also be able to provide them.

3.5.4 Designing Assessments that Align

What do we mean by *align*? The assessments you design should directly address student progress and/or demonstrate success at meeting a particular LO. This means that the task you are asking the students to do should match what the LO is saying they should be able to do. Here are some examples of assessments that do not quite align with the LO.

Topic-level learning objective
Level: (M) Making meaning. Students will be able to describe why we see different phases of the Moon from Earth.

The assessment task
Label the following diagram with the correct phase of the Moon:

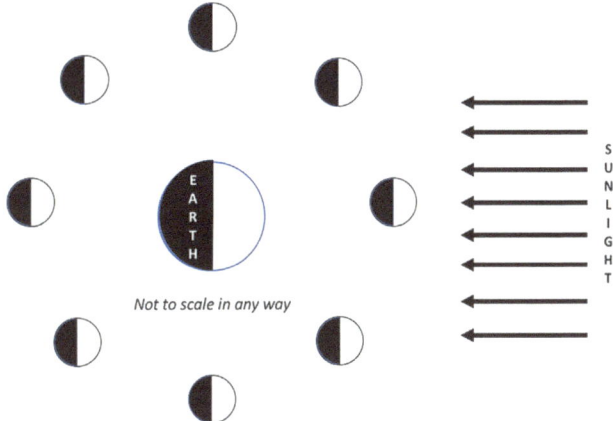

What is the mismatch?
Students are being asked to label a diagram that is assessing whether or not they can memorize a diagram, not if they understand *why* there are different phases of the Moon.

Discussion
What other task/assessment might work better to align with the LO? Labeling a diagram, which is a memorization task, is generally considered an (A) level LO. If the instructor needs an (M) level LO, then the assessment task needs to be more complex. What are your ideas?

---

[8] PhysPort-Peer Instruction, https://www.physport.org/methods/method.cfm?G=Peer_Instruction.

Learning objective	The assessment task
Level: (T) Transfer. Students will be able to compare and contrast the properties of different wavelengths of light.	Rank the following wavelengths of light in order of shortest wavelength to longest wavelength: gamma rays, infrared, visible, X-ray, microwave, ultraviolet, radio waves.

What is the mismatch?
For sure, a "ranking task" asks students to compare one thing to another; however, this assessment might only indicate if students have memorized a particular order of the properties.

Discussion
The instructor has labeled this LO as (T) so the assessment needs to be high level. What is a richer task to ask students to do where they can demonstrate some critical thinking on the properties of light? Or maybe the LO is not using the right wording and needs to be changed to be lower level. How might you change the LO so that the assessment above is more aligned with the task? Alternately, how might the question be elevated to be a more robust "compare-and-contrast" task?

Learning objective	The assessment task
Level: (A) Acquiring. Identify if a type of nebula absorbs, emits, and/or reflects light.	Describe the absorb, emit, and/or reflect processes for various types of nebulae.

What is the mismatch?
Consider if you were a student reviewing the LO for the unit, in preparation for a quiz or exam. How far would you go in understanding the mechanism? How might you study or practice? When confronted with the assessment task, would you be prepared for it?

Discussion
The assessment task is a much higher level than "identify." Students might focus on the properties or descriptions of nebulae in a memorization sense, and not "why or how" light will be reflected, absorbed, or reemitted. A lower-level objective might be: "Match the process (emit, absorb, and/or reflect) to the type of nebula." Alternately, the LO could be rewritten to reflect that students need to understand the how and why.

Frequently in course design workshops, the following concern is voiced, "But, if I am meant to do assessments/evaluations that match up to my LOs, in a very detailed and precise way...aren't I then *teaching to the test*?" A valid concern, and fear, particularly when some institutions are concerned with grade inflation. Suskie (2009) phrases a popular explanation well:

In a way, good assessment *is* teaching to the test. Assessment is part of a process that identifies what we want students to learn, provides them with good opportunities to learn those things, and then assesses whether they have learned those things (p. 12).

One of the responsibilities of an instructor is to help facilitate learning, giving students opportunities and resources to meet the LOs. Students ultimately are responsible for their own learning and motivation to learn, but you—the—instructor should expect to set up the environment, activities, and support for them to be successful. If you want them to reach higher and understand deeper, you have got to let them know how to do that and provide opportunities for practice, assessment, and feedback. This chapter aims to help you do that.

In the next task, you will research and brainstorm formative and summative assessments for your course-level, unit-level, and topic-level LOs. You will use formative assessments frequently with your topic-level and unit-level LOs and might group things together. As you brainstorm and design an assessment strategy, you might find that your LOs need to be changed. That is a good thing and means that you are thinking about the alignment and what you *really* want students to do. You might also notice that some of the assessment strategies feel like learning activities—you aren't imagining it! It is true that sometimes these are interchangeable, or better yet, they serve multiple purposes in learning. An assessment can be a learning activity. A learning activity can be an assessment. As previously mentioned, jot down off-task ideas on a sticky note and set aside for later. Don't get too stuck in the details if you aren't sure how to translate a face-to-face course assessment to an online course. You can come back later and work out the feasibility and implementation later.

Assessment and Feedback Guide.

Use these questions to examine your assessment strategy for each course-level, unit-level, and topic-level LO.

- *What is the purpose of the assessment?*

    You should be able to describe why the assessment is relevant, if it is more formative or summative, and how it will be timely to student learning or understanding progress.

- *Where and how do students get feedback, and what do they do with it? What do I do with feedback on student learning?*

    Remember that feedback might come from self-assessment or peers. If you cannot determine how or where feedback will be given or used when students participate in the assessment, then you might need to reconsider if the assessment is relevant and useful to the learning.

- *How does each assessment fit into the course-level LOs and course goals?*

    As you look at the hierarchy and nesting of the topic-, unit-, and course-level LOs, examine how the assessments weave into that framework or help hold it together. Do the assessments grow in rigor throughout the term?

- *If students have met the LO, what are some things they can show me? Compare to the products that will come from an assessment, do they match up?*

> If things do not match, you have options to rewrite the LO to be more relevant and/or redesign the assessment to be more relevant. Consider the ultimate purpose of the assessment to student learning and if you want to know progress, give an opportunity for feedback, or have it be a summative demonstration of their learning.

Task.

Use *Worksheet 7: Assessments and Feedback Strategies*[4] to describe a summative assessment for each course-level LO. Then, brainstorm assessment ideas for each unit-level and topic-level LO (or set of related LOs) and mark it as formative or summative. Identify places where you will get feedback on the progress of the learners and what you will do with that information. Then note where learners will get feedback and from whom (self, peer, instructor, system), and how it will help them learn.

Congratulations! You have completed Step 2!
Take a break from this chapter, worksheets, and notes. Return back in a few days and review your worksheets before heading into Step 3.

3.6 Step 3: Creating Learning Experiences

In Step 3: Creating Learning Experiences, you will consider how students are introduced to course content and acquire initial understandings, make meaning of ideas and concepts, transfer and deepen their understandings, and reflect on their learning. While exploring ideas and creating learning experiences, you might find that some of your assessments might very well be a learning activity, too. Likewise, learning experiences usually have built-in opportunities for formative assessment (feedback and reflection). Aim to be clear how everything works together and maps to the LOs.

3.6.1 Active Learning Primer

What is active learning? In general, an active learning approach to teaching will provide your students with varied opportunities to struggle with the content in front you and each other. The shared struggle or confusion is where the deeper learning can happen. They will confront misconceptions, integrate new understandings with prior knowledge, discuss what they think they know and what they do not know, and think critically about the course content. Brogt & Dokter (2019) emphasized why this type of engagement is critical:

> As revealed by the scholarship on teaching and learning, humans learn better through active engagement. In a class setting, this means "learning through

activities and/or discussion in class, as opposed to passively listening to an expert. It emphasizes higher-order thinking and often involves group work" (Freeman et al. 2014, pp. 8413–8414). These results have been rigorously tested across disciplines, including astronomy (e.g., Prather et al. 2009) and for science majors and nonmajors alike (e.g., Kober 2015; Freeman et al. 2014; Springer et al. 1999). As specific examples of active learning in astronomy teaching are discussed throughout the rest of this book, we will not reiterate them here, but there are many astronomy-specific active-learning tools that have been developed and are ready-to-use, either by individual students, pairs, or small groups (p. 1–7).

Fink (2013) illustrates his essential components to promoting active learning which "combine to form an enlarged and more holistic view of the topic—one that includes getting information and ideas as well as experiences and reflection" (p. 118).

The first component, getting information and ideas, asks instructors to figure out how students will be exposed to or discover the course content, either with primary sources or secondary sources (or both). Secondary sources are your lecture videos, textbook readings, or other media. Primary sources, however, allow students the opportunity to "read original sources and examine original data, that is, ideas and data that have not yet been fully analyzed and interpreted by others" (p. 120). Your decision on how to introduce course concepts to students will rely on your LOs and the general level of your course. Students in an introductory course might need more secondary sources at the beginning and can be introduced to level-appropriate primary sources for the (M) and (T) LOs. In the astronomy major and upper-level courses, you might consider having the majority of your content delivered via primary sources as you build up the professional skills of future scientists and researchers.

The second component, experiences, promotes constructing activities where students can make deeper meaning of the course content. What are the students actually doing or observing, and who are they working with? Initial meaning-making might involve online discussions in a forum or during a live session in a breakout room. There are also a variety of online simulations, applets, and citizen science projects waiting to have an instructional wrapper or activity built around the tool. Observing astronomers in action and using video chat to bring astronomers into the course experience are powerful and engaging strategies to help non-science-major students appreciate the field and develop a better grasp of the nature of science (Hickox & Gauthier 2014). Higher-level activities might involve analyzing or visualizing real astronomical data, projects involving communicating science to the public, collaboratively creating an observing plan, or being involved in a research project at your institution.

The third component for Fink's active learning philosophy involves reflection. As experts in their field, professors instinctively reflect on what they are learning and how it relates to, and might change, what they already know. Students however, need guidance and "time to reflect in order to decide what meaning to give" (p. 122) to the learning happening in the course. You can design these reflective moments by asking how new activities or concepts relate to other things in the course. A second type of reflection is prompting students to think about how they are learning, not just what they are learning. You might ask them to submit a lingering question after a lecture video or collaboratively annotate videos or articles with their questions. Tools like Hypothesis and VoiceThread allow this interaction with content either individually or in a collaborative asynchronous activity with peers. If one of your course goals focuses on the nature of science, you might ask students to submit journal entries throughout the term on how their worldview might be changing.

3.6.2 Choosing and Aligning Learning Activities

Learning activities can include passive tasks like reading, watching a video lecture, or listening to a podcast, as well as more active tasks like creating a concept map, discussing questions with peers, and collaborating on a group research project. Passive tasks fit well with LOs that are in the *acquire knowledge* category. The passive task of watching a video might be followed with a formative assessment of self-quizzing or outlining their notes. Students benefit from mini-lecture videos by reviewing and revisiting material multiple times until they get what they need from it.

For (M) and (T) LOs, you will need to design activities that will help students practice new skills, create new connections between course topics, and think critically and deeply about the content. These activities might require collaboration between students and facilitation by instructors and teaching assistants. When students work together to learn, they help each other figure out where confusion lingers and can grow deeper understandings together. Collaborative strategies also help to promote a learning community, where individual students might not feel as isolated while struggling. It's a challenge in fully online courses for students to feel and act like they are a part of a community. If you build in opportunities for students to work together and foster an inclusive environment, struggling students won't feel so alone while sitting solo at their computer.

The balance you find between active and passive learning in your course will greatly depend on your own teaching philosophy and how much you internalize a learner-centered viewpoint. It isn't all or nothing! There is a place for video lecture in your course. If you have not incorporated or experienced active learning in your online or face-to-face courses yet, then you might consider trying only one or two new things. In future course offerings, you can add to your active learning repertoire where appropriate for you, your content, and your learners.

What are the final steps on alignment? While you are choosing and designing the learning activities, you will be assessing their relevance and fit to the topic-level and unit-level LOs, and how the experiences build up to the course-level LOs. Consider the complexity and rigor of the learning activities against your AMT-categorized LOs as you assess your course design plan for alignment. You will also want to make sure the learning activities and experiences build students' knowledge and skills up to the level of the unit-level and course-level assessments. Alignment helps everything in the course feel more intentional, relevant to student learning, and helps the course flow from one section to the next.

Examples of astronomy-related activities and experiences that go beyond textbook reading, short lectures, and problem sets include

- Helping students explore and relate various online citizen science projects like CosmoQuest[9] and Zooniverse[10] to course topics. Learners can work with real data, contribute to the scientific community, and make the content they are learning less abstract and more authentic (see Chapter 8 for more).
- Designing an opportunity for students to discover course content through exploration of multi-wavelength imagery or tools like Chandra's openFITS[11] image-processing tool. Find out what students are curious about and give them a way to pursue answers.
- Allowing students to struggle with the course topics in pairs or small teams through higher-level polling questions or lecture-tutorial-style worksheets.
- Developing open-ended questions for students to answer using online animations or simulations like PhET[12] (see Chapter 5 for more).
- Constructing multistep problems for small group exercises that include regular stopping points to check in with teammates.
- Using a platform like WorldWide Telescope[13] for students to create presentations about our multi-wavelength universe set in the backdrop of an interactive observable universe (see Chapter 9 for more).

3.6.3 Resources

The following resources are rich with activity ideas. Each addresses course design and learner-centered teaching methods a bit differently, so you will get multiple perspectives on the tasks and content in this chapter. Your own department might have a budget for teaching resources, and these would be a high-impact addition to a bookshelf.

Learning Activities and Experiences Resources:

- University of Florida's Center for Instructional Technology and Training "Active Learning in Online Courses" resource http://citt.ufl.edu/online-teaching-resources/activelearning/
- UC Davis' public course "Teaching Online" is full of resources and ideas for engaging online learners. https://canvas.ucdavis.edu/courses/34528/pages/learning-activities-and-active-learning-online?module_item_id=4973
- Barkley et al.'s (2014) *Collaborative Learning Techniques: A Handbook for College Faculty* has sections on implementing collaborative learning as well as example activities with guides.
- Angelo & Cross' (1993) *Classroom Assessment Techniques: A Handbook for College Teachers* is a double-duty guide book. Sometimes, formative assessment can also be a learning activity.

---

[9] https://cosmoquest.org/x/.
[10] https://www.zooniverse.org/.
[11] http://chandra.harvard.edu/photo/openFITS/.
[12] https://phet.colorado.edu/.
[13] http://www.worldwidetelescope.org/webclient/.

- Barkley & Major's (2016) *Learning Assessment Techniques: A Handbook for College Faculty* is another dual-purpose book. The learning assessment strategies in the guide can be used as a learning activity.
- Barkley's (2010) *Student Engagement Techniques: A Handbook for College Faculty* is like her LATs book with a section on conceptual frameworks, tips and strategies, and implementation guides for 50 different activities.
- Center for Astronomy Education classroom materials and workshops: https://astronomy101.jpl.nasa.gov/materials/.
- Design your own lecture-tutorial-style activities with this guide from the Science Education Resource Center at Carleton College: https://serc.carleton.edu/NAGTWorkshops/teaching_methods/lecture_tutorials/index.html.
- Online physics simulations from PhET: https://phet.colorado.edu/.
- Online astronomy simulations from the Nebraska Astronomy Applet Project: https://astro.unl.edu/naap/.
- Open-source night-sky visualization software, Stellarium: https://stellarium.org/.

Learning Activities Guide.

Questions to ask for each LO and/or course objective:
- *Does the activity address the level of the LO (acquire, making meaning, transfer)?*
  By participating in the activity, what will students be doing that will help them make progress toward the LO? Is it at the right level?
- *Is there flexibility in the activity for students at different skill levels?*
  Will those who need more time to show progress feel left out or become anxious? Will those who move quicker get bored or have something advanced they can do?
- *Is there an opportunity, and is it feasible, for students to collaborate and learn from each other?*
  Collaborative activities and learning experiences have built-in formative assessment and feedback as students discuss a problem, brainstorm a strategy, or create something together. They can help each other uncover misconceptions and gain clarity through alternate explanations and viewpoints.
- *How are students held accountable for participating or being engaged in the activity?*
  Certainly you would like to see students be self-motivated and engaged, but much of the time there needs to be some sort of accountability built into the activity. It might be they hand in a worksheet, complete a quick reflection, or get credit for showing up and being on task.

Task.

Use *Worksheet 8: Learning Activities and Experiences*[4] to brainstorm and outline learning activities and experiences for each LO or group of LOs. You will check for alignment, rigor, flexibility, collaboration opportunities, and individual accountability.

Congratulations! You have completed Step 3!
Once again, I suggest walking away from the chapter, worksheets, and notes. Return back in a few days and review your worksheets before heading into Step 4.

3.7 Step 4: Putting It All Together

In Step 4: Putting It All Together, you will examine the overall alignment of your course plan for your one course goal by reviewing all of your LOs, assessments, feedback, and activities. You will consider the flow of the course and connections between the units. Finally, you will reflect on how you will communicate your course plan to students and colleagues.

3.7.1 Checking for Alignment

At this "checking for alignment" stage, the iterative, reflective, and flexible nature of this course design process becomes obvious. You might go back and change a course goal or LO when you have another look at the assessments and activities. You might determine you have too many unit-level LOs or that your topic-level LOs ballooned to an unreasonable number and you need to pare it down. You might identify opportunities where you can merge topics or activities to get students thinking at a higher level more efficiently. You could even realize one of your "big ideas" or course goals is not attainable or realistic.

This adjustment will continue as you teach. However, by the time the course starts, your course map should describe how the learning activities and experiences help the students make progress toward the LOs, which connect with relevant and appropriate assessment and feedback plans. That is alignment.

Course Alignment Guide.

Questions to ask yourself as you review your course map:
- Are your course goals and course-level LOs written in student-friendly language?
- Do your course goals describe big-picture ideas and what you hope your students will remember most from the course?
- Do your course-level LOs have the proper level of rigor and speak toward the final assessments or products of the course?
- Do your topic-level and unit-level LOs make relevant connections to your course-level LOs? How do they relate to each other or nest within each other?
- Does the summative assessment plan clearly describe what students will do to show they have met your course-level LOs?
- What are the opportunities for formative assessment and instructor feedback as students make progress toward the course-level LOs?
- Are there specific places for students to figure out what they know and do not yet know, and then how to act on that information?
- Do your learning activities and experiences give students a chance to learn-while-doing or practice at high enough levels to meet your summative assessment expectations?

> *Task.*
>
> Use *Worksheet 9: Course Plan*[4] and the questions in the Course Alignment Guide to review your work on one of your course goals.

3.7.2 Finding the Flow

Throughout this course design journey, you have been asked to think about the alignment of the course goals, LOs, assessments, and activities at different scales. The smallest scale is at the topic level or the class-period level, which nests or weaves into the units that build up to the ultimate course-level LOs. This can be described as flow. The flow of a course can be viewed as the connective tissue that holds the entirety of the course together into one cohesive learning experience.

Flow also needs to exist at the microlevel of how a given class period or activity flows to the next throughout the week. It might seem obvious to you how watching a video lecture on Monday night is linked to group problem solving activities during the live session on Tuesday; however, students need to be told how these activities are dependent on each other. Being transparent and specific with students will help them find relevance and motivation. Give students opportunities to discover how last week's class topics relate to the assignment over the weekend and will build up to next week's more rigorous tasks. Build the connective tissue into all aspects of your course. You might do this with the sequence of activities, frequent reflection and assessment, or asking students to talk with peers about the connections in an online discussion forum.

In this mock workshop, you were asked to focus on one course goal or course-level LO. In reality, you might have three to five course-level LOs where each breaks out into a few units. Your course structure might be tightly organized into an outline with nested bulleted lists. Or, it might be less of a hierarchy and more like a concept map of related ideas. Maybe your course makes use of various themes that weave together throughout the term. Whatever the style, it is your job to find clarity in the course design and clear up any tangles. If you have trouble determining the structure and flow of your course, focus on how you would describe the themes and topics of the course to someone else who might not be an expert in the field.

> *Flow guide.*
>
> Questions to ask yourself as you review your course plan:
> - How do the class-period or topic-level activities build up to and prepare students for more complex unit-level activities and assessments, and even higher to the course-level LOs?
> - In what ways are you communicating to students the relevance between activities and assessments?

- How are students finding and reflecting on the connections between the units in your course?
- Does your course plan or map communicate the structure and flow of your course in an obvious and transparent way?

Task.

Complete *Worksheet 10: Reflecting on Flow*[4] to answer the Flow Guide questions.

3.7.3 Communicating Your Course Plan

The course plan or map you create for yourself might look like a lesson plan with lots of details, to do lists, and planning notes. Though you would not share this level of detail with your students, you might share this course plan with graduate teaching assistants or undergraduate learning assistants. It is a living document that can be updated and shifted around throughout the term, particularly if you ask students for early and mid-course feedback.

The course plan, map, or outline you do share with students should not be intimidating and overwhelming. An approachable method is to share the high-level course goals, course-level LOs, and unit-level LOs in the syllabus or a well-organized page in your course site. You can leave the class- or topic-level LOs for when they will encounter that material. In an effort to be transparent with your course design, consider listing or describing where progress is checked through formative assessments and clearly mark where summative assessments occur. Describing or outlining the activities, experiences, and assessments relevant to each unit- or course-level LO helps show students the path to success. You might create a page at the beginning of the week/module for students or record a 2-minute video where you describe how everything in the module weaves together.

You might not want to use the word "alignment" in your communications to students, but you can articulate the relevance of the learning activities to the assessments. It can motivate and engage students when they understand why they are doing something and how it will help them succeed. A well-communicated course plan can help with buy-in, particularly in courses where active and collaborative learning are the focus.

3.8 Resources for Teaching Online

Teaching online is both a similar and a drastically different endeavor from teaching students in a face-to-face classroom. Thankfully, over the past two decades higher education instructors, education researchers, faculty developers, and learning designers have implemented and tested a variety of strategies to figure out what can work best. It's not just about effective and aligned course design, it's also about the learning environment that you create and foster. How can you promote inclusivity and a sense of belonging? What strategies will help make your online course inviting, engaging, and feel like a community? How will you get to know your students as people, and help them get to know (and help) each other? The resources below have been intentionally gathered to give you multiple perspectives on online teaching. Whether it's a small undergraduate course or a worldwide massively open online course (MOOC), you should be able to find strategies in the resources below that will work with your teaching philosophy.

Resources:
- Flower Darby's lens on James Lang's *Small Teaching: Everyday Lessons From the Science of Learning* is summarized in the Inside Higher Ed article, "Small Teaching[14] Online." Read the interview and then get their new book "Small Teaching Online: Applying Learning Science in Online Classes"[15] (Darby & Lang 2019).
- Review strategies from Michelle Pacansky-Brock for How to Humanize Your Online Course[16] and help your students be engaged, feel valued, and create community. What is meant by "humanizing"? Check out this EdSurge article by the same author, How to Keep the Human Element in Online Classes.[17]
- The Innovative Learning team at Purdue University has put together an in-depth guide to online teaching and learning. Check out their PoRTAL (Purdue Repository for online Teaching and Learning) resource.[18]
- In *How to Be a Better Online Teacher ADVICE GUIDE*[19] (Darby) in the Chronicle of Higher Education, you will find "advice on how to make your online pedagogy as effective and satisfying as the in-person version" (Darby 2019).
- UC Davis' open resource can help you think about Being Present in Your Online Course.[20]
- Melissa Wehler shares Five Ways to Build Community in Online Classrooms[21] through the Online Education section of Faculty Focus.[22]

3.9 Conclusion

You have learned about and tried the main steps of an integrated course design method (Figure 3.1) by
- developing course goals and articulating specific LOs,
- creating a formative assessment and feedback strategy to evaluate student progress toward the LOs,

---

[14] www.insidehighered.com/digital-learning/article/2019/06/26/bringing-small-teaching-online-classroom
[15] www.wiley.com/en-us/Small+Teaching+Online%3A+Applying+Learning+Science+in+Online+Classes-p-9781119619093
[16] https://brocansky.com/humanizing-infographic
[17] www.edsurge.com/news/2016-04-27-how-to-keep-the-human-element-in-online-classes
[18] www.purdue.edu/innovativelearning/supporting-instruction/portal/
[19] www.chronicle.com/interactives/advice-online-teaching
[20] https://canvas.ucdavis.edu/courses/34528/pages/being-present-in-your-online-course
[21] www.facultyfocus.com/articles/online-education/five-ways-to-build-community-in-online-classrooms/
[22] www.facultyfocus.com/topic/articles/online-education/

- creating a summative assessment strategy to evaluate student success with the high-level LOs,
- designing learning activities and experiences that help the student learn and practice, and
- developing a communication plan.

This was just a start, a taste of being structured and decidedly intentional in designing a course or learning experience. The next step is to branch out and work with all of your course goals, not just the one you used here. It can be an overwhelming and time-consuming process, so give yourself plenty of time before the term.

Please consider the advice from the beginning of the chapter:
- Seek out interested colleagues who can either work along you on their course, or can give you constructive feedback.
- Find your teaching and learning center or those at your institution who help faculty with teaching and learning in online course environments.

Thank you for joining me on this course design journey, in a format that is very much different from a live in-person workshop or consultation. I don't get to see how you (readers) are learning and progressing through the steps, and then hearing about how it went during the term. Wishing you the best in your teaching and course design futures!

Please visit the Course Design Workshop companion site where you can find more resources, worksheets, and reflections!

Reflect.

Consider answering the following questions:
1. How does your experience of using an integrated course design process compare to what you have done previously? Refer back to your first reflection in this chapter.
2. What did you find most useful about this experience?
3. What was most engaging for you?
4. What did you find most challenging?
5. Which aspects of this integrated course design process might you incorporate into your regular course and learning design practices?
6. What are your lingering questions?

I'd love to know your answers to these reflection questions! Please consider visiting the Course Design Workshop site (https://sites.google.com/view/cdw-companion) to submit your answers and lingering questions. Knowing what has engaged or motivated you in the course design process, as well as hearing your challenges and lingering questions will help me improve this chapter on the next iteration. This is my way of getting some formative assessment!

Acknowledgments

I would like to show my appreciation for my teaching and learning colleagues at Dartmouth College. If it were not for them, this chapter would not really exist. The steps and ideas here come from many conversations, many faculty workshops and institutes, and many resources and articles shared on our Slack channel. An appreciative thank you to the wonderful faculty and students at Dartmouth College who are an inspiration to my daily work. A special shout out to Prue Merton who helped me through this writing project and gave me fabulous, useful, thoughtful, and encouraging feedback along the way. Thank you, Prue! I'd also like to thank Sanlyn Buxner for this opportunity. Lastly, I am sending a big wave through the air to all my astronomy education colleagues, whom I miss working and networking with at super-cold winter AAS meetings.

References

Anderson, L., Krathwohl, D., & Bloom, B. 2001, A Taxonomy for Learning, Teaching, and Assessing: A Revision of Bloom's Taxonomy of Educational Objectives (New York: Longman)

Angelo, T., & Cross, K. 1993, Classroom Assessment Techniques: A Handbook for College Teachers (2nd ed.; San Francisco, CA: Jossey-Bass)

Bailey, J. M. 2019, in Astronomy Education, Volume 1: Evidence Based Instruction for Introductory Courses, Evidence based instruction for introductory courses, ed. C. Impey, & S. Buxner (Bristol: IOP Publishing), 10-1

Barkley, E. F. 2010, Student Engagement Techniques: A Handbook for College Faculty (San Francisco, CA: Jossey-Bass)

Barkley, E., & Major, C. 2016, Learning Assessment Techniques: A Handbook for College Faculty (San Francisco, CA: Jossey-Bass)

Barkley, E. F., Cross, K. P., & Major, C. H. 2014, Collaborative Learning Techniques: A Handbook for College Faculty (San Francisco, CA: Jossey-Bass)

Brogt, E., & Dokter, E. 2019, in Astronomy Education, Volume 1: Evidence Based Instruction for Introductory Courses, Evidence based instruction for introductory courses, ed. C. Impey, & S. Buxner (Bristol: IOP Publishing), 1-1

Carl Wieman Science Education Initiative (CWSEI) & The Science Education Initiative at the University of Colorado (CU-SEI) 2017, Clicker Resource Guide, http://www.cwsei.ubc.ca/resources/files/Clicker_guide_CWSEI_CU-SEI.pdf

Darby, F. 2019, Small Teaching Online: Applying Learning Science in Online Classes, https://teachinginhighered.com/podcast/small-teaching-online/, https://lacol.net/small-teaching-online/

Darby, F., & Lang, J. 2019, Small Teaching Online; Applying Learning Science in Online Classes (San Francisco, CA: Jossey-Bass)

Fink, L. D. 2013, Creating Significant Learning Experiences: An Integrated Approach to Designing College Courses (San Francisco, CA: Jossey-Bass)

Fink, L. D. 2005a, Idea Paper #42 Integrated Course Design, https://www.ideaedu.org/Portals/0/Uploads/Documents/IDEA%20Papers/IDEA%20Papers/Idea_Paper_42.pdf

Fink, L. D. 2005b, A Self-Directed Guide to Designing Courses for Significant Learning, https://www.deefinkandassociates.com/GuidetoCourseDesignAug05.pdf

Freeman, S., Eddy, S. L., McDonough, M., et al. 2014, PNAS, 111, 8410

Haave, N. 2014, Six Questions That Will Bring Your Teaching Philosophy into Focus, https://www.facultyfocus.com/articles/philosophy-of-teaching/six-questions-will-bring-teaching-philosophy-focus/

Hickox, R., & Gauthier, A. 2014, in 224th Meeting of the American Astronomical Society, Poster Session (Washington, DC: AAS)

Kober, N. 2015, Reaching Students: What Research Says about Effective Instruction in Undergraduate Science and Engineering (Washington, DC: National Academies Press)

Mazur, E. 1997, Peer Instruction: A User's Manual (Upper Saddle River, NJ: Prentice Hall)

Nicol, D. J., & Macfarlane-Dick, D. 2006, Studies in Higher Education, 31, 199

Prather, E. E., Rudolph, A. L., Brissenden, G., & Schlingman, W. M. 2009, AmJPh, 77, 320

Prather, E. E., & Wallace, C. S. 2019, in Astronomy Education, Volume 1: Evidence Based Instruction for Introductory Courses, Evidence based instruction for introductory courses, ed. C. Impey, & S. Buxner (Bristol: IOP Publishing), 3-1

Springer, L., Stanne, M. E., & Donovan, S. S. 1999, Review of Educational Research, 69, 21

Suskie, L. 2009, Assessing Student Learning: A Common Sense Guide (Bolton, MA: Anker Pub. Co.)

Weimer, M. 2012, Five Characteristics of Learner-Centered Teaching, https://www.teachingprofessor.com/topics/teaching-strategies/active-learning/five-characteristics-of-learner-centered-teaching/

Weimer, M. 2013, Two Activities that Influence the Climate for Learning, https://www.teachingprofessor.com/topics/for-those-who-teach/two-activities-that-influence-the-climate-for-learning/

Wiggins, G. P., & McTighe, J. 2008, Understanding by Design (Alexandria, VA: Association for Supervision and Curriculum Development)

Wiggins, G. P., & McTighe, J. 2013, Essential Questions (Alexandria, VA: Association for Supervision and Curriculum Development)

Chapter 4

Astronomy Education in Virtual Worlds and Virtual Reality

Anthony Crider

The night sky is a beautiful thing to the human eye. It can seem almost sacrilegious to suggest that humans might instead experience the universe by strapping tiny electronic screens to their heads, shutting out the real world, and viewing a flawed simulation of it. Nevertheless, in this chapter, we will suggest exactly that. We will begin by reviewing astronomy education that took place in *virtual worlds* (VWs) that became popular in the mid-2000s, most notably Second Life. We will then examine the current options for astronomy education with *virtual reality* (VR) technology, such as the Oculus Rift, that became available in the mid-2010s. Finally, we will speculate what might be done in *virtual reality virtual worlds* (VRVWs) in the decade ahead as we consider what kind of VR experiences we should encourage to enhance and promote astronomy education in both formal and informal settings.

Chapter Objectives

By the end of this chapter, readers will be able to:
- Identify the distinction between virtual worlds, virtual reality, and virtual reality virtual worlds.
- Describe the basic history of virtual worlds, virtual reality, and virtual reality virtual worlds.
- Describe methods instructors have used with virtual worlds and virtual reality in astronomy education.
- Understand issues regarding how virtual reality and virtual worlds might be used in astronomy education in the future.
- Find resources for using virtual reality and virtual worlds to teach astronomy.

4.1 Introduction

The value of a simulated starfield is far from new. Orreries have existed since antiquity. These mechanical devices evolved in their complexity over centuries. It was less than a century ago that we attempted to simulate the stars and planets using light. In 1923, the Carl Zeiss company created the Mark I planetarium, the first to use lenses and light bulbs to project stars and planets onto a white dome. Six decades later, Evans and Sutherland created a laser-based projector that could cover the dome with any starfield (or lines) which later gave way to full-dome digital projectors that could project any image. These could display virtual environments that surrounded the audience, providing an immersive and moving perspective of computer-rendered animations (Yu 2005).

While the real night sky is indeed quite beautiful, there are many, many reasons we might want to have virtual skies. Nearly all formal education occurs in schools during the day. Many nights, weather frequently makes sky viewing impossible. Even under perfect skies, the most basic celestial motions happen so slowly they are nearly imperceptible to humans. Furthermore, viewing the sky from extreme locations (e.g., the North Pole, the Equator) can offer a more simplistic view of the sky (with stars moving parallel or perpendicular to the horizon) which can be more readily comprehended by astronomy students.

These advances in technologies, from orrery to planetarium to digital projector, improved both the realism and the flexibility of the simulations. The next step in this path is virtual reality, or VR. With the recent debut of commercially available VR headsets, there are new opportunities for astronomy education. To consider how virtual reality might impact formal and informal astronomy education in the coming decades, we begin by reviewing a closely related technology: virtual worlds.

4.2 Astronomy Education in Virtual Worlds

4.2.1 What Are Virtual Worlds?

Many people have been introduced to the concepts of both virtual reality and virtual worlds through science fiction. In his 1992 novel "Snow Crash," Neal Stephenson described several technologies in a post-apocalyptic near future, including a virtual globe that one could spin around, zooming in and out to see places at multiple scales. In his book this software was called *Earth*; today we call this *Google Earth*. Stephenson also described software, the Metaverse, in which many users would simultaneously navigate their virtual avatar bodies along a virtual street with virtual buildings. Years later, "Ready Player One," a 2011 novel by Ernest Cline and its 2018 movie adaption by Steven Spielberg, made the concepts of virtual reality and virtual worlds even more mainstream. The live-action humans in the movie don virtual reality headsets, gloves, and even body suits to control digital versions of themselves, known as avatars. These avatars could fight, dance, and explore the virtual world of OASIS.

In the first issue of the "Journal of Virtual Worlds Research," a few scholars offered their definition for this emerging form of software. Mark Bell (2008) wrote that a virtual world was a "synchronous, persistent network of people, represented as avatars, facilitated by networked computers." In that same issue, Ralph Schroeder (2008) argued that "[v]irtual worlds are persistent virtual environments in which people experience others as being there with them—and where they can interact with them." Neither Bell nor Schroeder suggested that a virtual world must be in 3D or experienced using virtual reality head-mounted displays (HMDs) akin to the ones necessary to experience the Metaverse or OASIS in science fiction.

The vast majority of virtual worlds to date have been video games. Many of these are massively-multiplayer online role-playing games (MMORPGs) such as Ultima Online (1997), Everquest (1999), and then World of Warcraft (2004). These worlds allowed for socialization and role-playing while still progressing users through a series of quests that constituted the game. Second Life, however, was not a game. Unlike the many MMORPGs which offered a full world for users to compete in, Second Life offered an empty world for users to create and play in. Launched in 2003, Second Life was an attempt by Philip Rosedale and his company, Linden Lab, to make the fictional Metaverse described in Snow Crash into a reality

4.2.2 How Has Astronomy Education Been Done in Virtual Worlds?

The launch of Second Life occurred at a time when educators were beginning to take games seriously. Games for Change was founded in 2004 and in 2010 the first Serious Play Conference was held. At the first Game+Learning+Society meeting in Wisconsin in 2005, Conklin (2007) proposed 101 uses for Second Life in higher education. While she suggested several unique ways to use Second Life as a thing to be studied using various disciplinary lenses (e.g., economics of Second Life, sociology of Second Life), a great deal of education has instead used Second Life as the medium for sharing disciplinary content. This often happens via lectures in virtual lecture halls or exhibits in virtual museums. The 101st thing listed, and one of the first examples of astronomy education in this virtual world, was my own, The Second Life Planetarium. Originally located on a private parcel of virtual land, this was eventually moved to a new Linden-owned campus, and then to the International Spaceflight Museum (ISM; Figure 4.1). The former was a brief, first foray by Linden Lab to encourage education. The latter was a group of space exploration enthusiasts that built a virtual rocket museum filled with virtual exhibits designed for visitors to virtually walk through before exiting through a virtual gift shop.

The ISM became a model for real-life education within Second Life and eventually served as the cornerstone for the SciLands, an archipelago of virtual islands devoted to science education. Member institutions of the SciLands included NASA, NOAA, NPR's Science Friday, the Exploratorium in San Francisco, the National Physical Laboratory in the UK, and multiple universities. Each of these places paid Linden Lab to lease a server which would store the data and run the software necessary to create a virtual island.

The majority of the SciLands exhibits were dedicated to space and earth science. Some, like the Exploratorium, emulated the ISM and created virtual science

Figure 4.1. The Viking Lander display at the International Spaceflight Museum within Second Life.

museums. Some parts of the SciLands tried to create virtual space destinations difficult or impossible to visit, including a Martian landscape on NASA's virtual island, a recreation of the Meyer–Womble Observatory by the University of Denver, and my own simulation of the Apollo 11 landing site.

Driven by this early interest in virtual world education, NASA put out a request for proposals to develop an educational, space-based video game (Laughlin 2008). This wouldn't just be a single-player game, though; they wanted a massively-multiplayer online game and said as much in their RFP:

> A NASA-based MMO game built with the goal of engaging young people can enhance STEM education by using NASA-based content that draws and holds their attention with fun and challenging game play. The power of games as educational tools is rapidly gaining recognition. Innovative university faculty are already holding classes and taking field trips to synthetic worlds like World of Warcraft® and Second Life®.

NASA eventually selected The Army Game Studio (developer of the first-person-shooter, America's Army) to develop Moonbase Alpha, which came out in 2010 (Figure 4.2). In this game, players take the role of a NASA astronaut on a futuristic moonbase and must repair it when a meteor strikes. The game did become popular due to its text-to-speech function, which users hacked to make the astronauts sing to one another.

Figure 4.2. NASA's Moonbase Alpha game. Courtesy of NASA.

While it is much more common to use virtual worlds as a small piece of an astronomy course, "At Play in the Cosmos" (Bary & Frank 2018) was developed by W. W. Norton to be the dominant piece of the course. This mission-based game includes a companion textbook covering standard "Astronomy 101" content. This isn't technically a "virtual world" (since it is not networking people). Still, it has many things in common with Moonbase Alpha. Both games won awards for their innovative approach. However, these rely primarily on accomplishing quests to beat the game and not all gamers are motivated by such quests.

In a seminal study of motivations among 30,000 MMORPG players, Nick Yee (2006a) found three main components: achievement (advancement, mechanics, competition), social (socializing, relationship, teamwork), and immersion (discovery, role-playing, customization, and escapism). Statistical differences existed between the motivations reported for male versus female gamers, with male gamers being somewhat more motivated by achievement factors and female gamers being somewhat more motivated by immersion factors (Yee 2006b). Unlike the achievement-centric astronomy games mentioned above, Second Life allowed for socializing, relationships, teamwork, and customization. As a result, Second Life has a 61% female population compared to sci-fi MMOs (11%) that are similar to Moonbase Alpha (Yee 2017).

The success of another virtual world also showed that building games can draw a very broad audience. With 144 million copies sold, Minecraft has become an international phenomenon and is arguably the largest social virtual world (Crider & Torrez-Riley 2017). While it began as merely a construction game, it added elements of survival, and eventually the ability to socialize with other players (Duncan 2011).

These added elements allowed Minecraft to address the diverse motivations (achievement, social, and immersion) of many potential players.

The science outreach done within Second Life served as an informal education opportunity for users that had already bought into the concept of a virtual world. Second Life residents gathered together to question science lecturers, to watch live-streams of eclipses, and to listen to broadcasts of NPR's Science Friday. If they were primarily interested in learning the content itself, the could have view higher-quality videos on nasa.gov or YouTube. However, for these people, the simulated museums and college campuses met the expectations of what an "educational experience" looks like. In at least one case, audience members even lined themselves up outside of a virtual planetarium to wait and watch the show (Figure 4.3).

The fact that Second Life was a virtual world populated with thousands of real users led to an interesting hybrid of formal and informal learning. Formal astronomy education in the United States is predominantly done in large classes that use lectures to convey content and multiple-choice tests to assess learning. In their book, "Strategies for Astro 101", Slater & Adams (2003) suggest dedicating up to 30% of class time on non-lecture group work that in some cases might result in portfolio assessment such as a three-ring binder of classroom materials or a multimedia tour. Green (2003) offers details on how why such group work is beneficial and how to form them. Students taking a university class for credit could build a 3D astronomy-themed displays and then have an authentic audience for it, that is other Second Life residents. End-of-semester projects were used for

Figure 4.3. Residents of Second Life form a line to enter The Second Life Planetarium.

assessment in multiple astronomy classes using Second Life (Crider 2006; Gauthier & Impey 2008). Compared to other sciences, many astronomical objects were *relatively* easy to create and render using the Second Life tools. Planets can be rendered as spheres and the blackness of space, or even a starfield, is an easy backdrop for many exhibits when compared to a building a chemistry lab or a wildlife scene. These projects sit near the top of Bloom's Taxonomy (Bloom 1956; Krathwohl 2002), which lists several levels at which students can cognitively process a topic with increasing complexity: remembering understanding, applying, analyzing, evaluating, and creating.

The very first astronomy student projects were planetarium shows, with a narration and slides generated by students and played within a scripted Second Life planetarium. Later, students began making small museum-like exhibits based off of curriculum. Some student groups researched and built additions to existing Second Life displays (e.g., a seismometer for the Apollo 11 simulation). At other universities, students built pieces for a special "History of the Earth and Life on Earth" museum (Figure 4.4).

During one semester when "An Inconvenient Truth" by Al Gore was the campus common reading, all of the groups built projects that connected to climate change. One student group constructed a 3D-house with green gas oozing from the various appliances. They had done research and calculations to make the amount of green

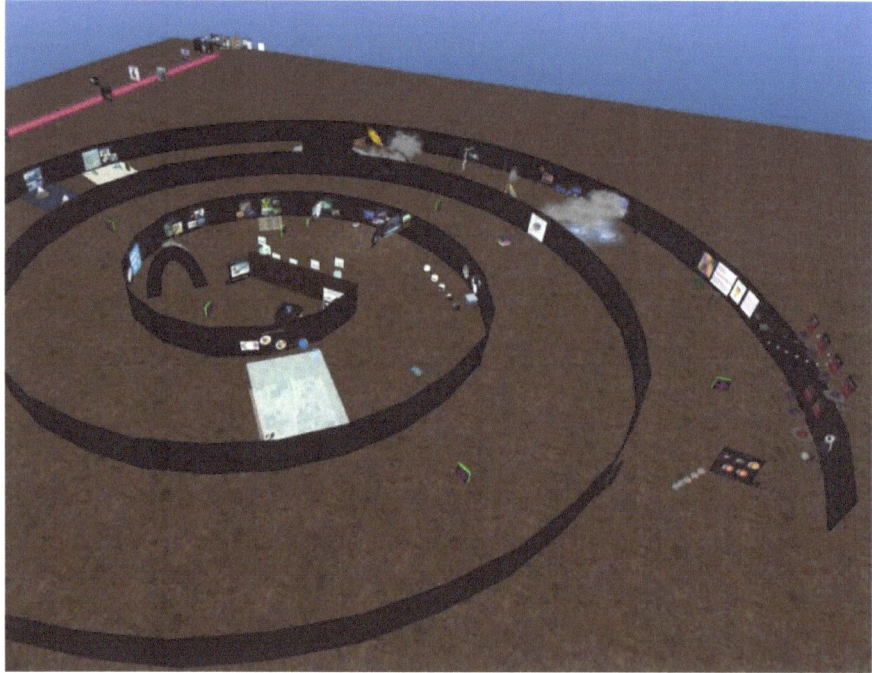

Figure 4.4. Student-created displays along the "History of the Earth and Life on Earth" within Second Life.

gas proportional to the amount of greenhouse gas contribution the resulted from the use of the appliance (Figure 4.5).

4.2.3 Getting Started in Virtual Worlds

Over fifteen years after its launch, the Second Life platform continues to be the most popular virtual world for education. At its peak in 2009, over 60,000 people would be concurrently logged in (Tateru 2012). While today that number has dropped almost in half, it is still much more than any other competing social virtual world (Crider & Torrez-Riley 2017) with the exception of the very low-resolution but extremely popular Minecraft. One can create a free account, pick a starting avatar, and select a character name at secondlife.com. The avatar can be easily changed later. In fact, much of Second Life revolves around character customization. After downloading the Second Life client to your computer, you can log in and begin. Avatars begin in an orientation area that teaches new users how to move around.

Since Second Life is a social virtual world, one should probably start by become familiar with the controls in one of the more popular destinations (found by clicking the Destinations button). Examples include dance clubs, skill gaming regions (which replaced illegal gambling regions), and role-play areas which are often fantasy or sci-fi themed. It can be help for new users to start with a destination that they have some familiarity with in real life (also called "RL" in the "SL" community).

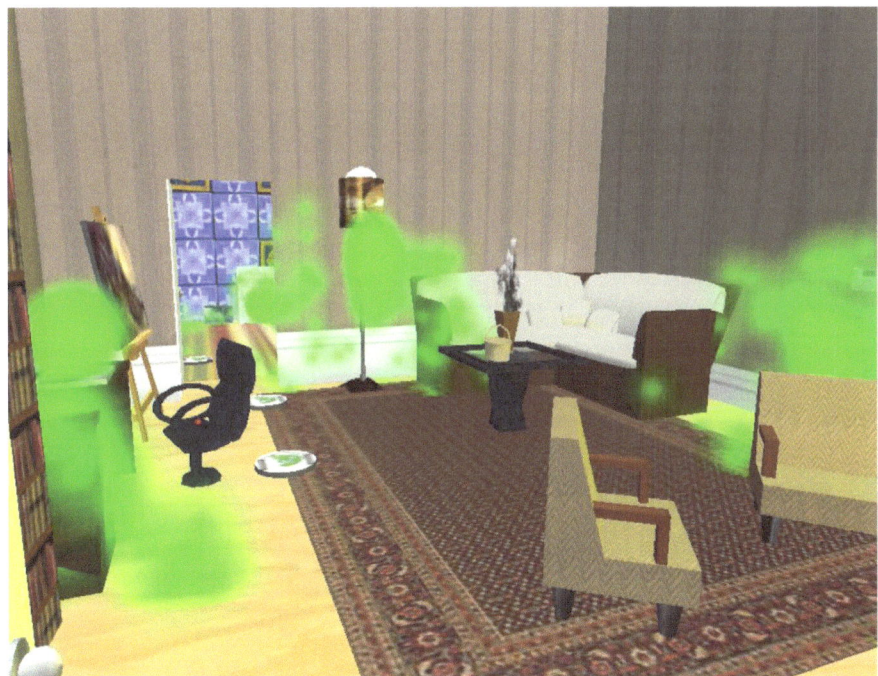

Figure 4.5. The student-created "Greenhouse Gas House" exhibit within Second Life.

Educators new to Second Life can find mentors in one of the many education groups in it. Some groups hold education conferences within Second Life, most notably the Virtual Worlds Best Practices in Education conference (https://vwbpe.org/) that has been held annually since 2007. Some groups serve primarily as a mailing list to notify educators of various events (e.g., "Real Life Education in Second Life" created by Linden Lab).

One can explore existing exhibits in Second Life with a free avatar but to create a new educational experience, one needs to own (or rent) a land parcel. In reality, of course, one is paying for server space and use of the code. With a premium membership ($10 USD per month), users receive a small virtual house in which they can build things as well as the right to rent additional land on the Second Life mainland. However, to have more control over what buildings are next door, many educators opt for private virtual islands. This is what led to the creation of the SciLands mentioned above. At the time of writing, a small private island (able to host 20 users simultaneously) costs $110 per month plus a $150 set-up fee. The more standard private island size (able to host 100 concurrent users) costs $250 per month after a $350 set-up fee.

It is not too difficult to build new, three-dimensional objects within Second Life using the tools in the client itself, so long as one is only applying textures to standard geometries (e.g., spheres, cylinders, blocks) called "prims." Given that so much of astronomy involves spheres, it is possible to have new students build a small solar system in Second Life with minimal instruction. The Linden Scripting Language (LSL) used to create interactivity within Second Life is not well-suited for teaching students since it is too complex for non-STEM majors and not similar enough to standard programming languages to be useful to STEM majors. That said, small scripts (e.g., to make an object rotate, to give an object a label) can be "dragged-and-dropped" onto virtual objects making it easy for non-programming students to add functionality to Second Life prims. Post-class assessments have shown that students appreciate the opportunity to create in virtual worlds. (Jarmon et al. 2009; Schiller 2009). Pre- and post-class quizzes also have shown that students using Second Life to learn science, specifically chemistry lab techniques, have similar learning gains for those learning in a regular lab (Winkelmann et al. 2017). This should not be surprising since well-designed online activities can actually lead to higher learning scores than "real" laboratories, as Finkelstein et al. (2005) found for electronics laboratory simulations.

Given the easy-of-use for making novel, educational experiences in Second Life but the relatively high-cost to use Linden Lab servers, some educators have opted to use the open-source version, OpenSimulator. These can be accessed with third-party Second Life clients, such as the popular Firestorm. Small businesses such as Kitely offer private islands (that is, server space) for as little at $15, over ten times less. For formal education setting in which the students don't need to interact with the outside world, this is ideal. For informal education, where one wants to draw an audience, the large user base of Second Life may outweigh the cost savings of using OpenSimulator.

4.3 Astronomy Education in Virtual Reality

4.3.1 What Is a Virtual Reality?

In both Snow Crash and Ready Player One, users explored the virtual worlds (existing only in software) using virtual reality hardware. In a time when no VR headsets were available, before Google Cardboard or the Oculus Rift, Jonathan Steuer (1992) laid out a psychology-based (rather than a technology-based) definition of virtual reality that went beyond the "goggles 'n' gloves" and instead could be any "real or simulated environment in which a perceiver experiences telepresence." This broader definition, which might include remotely operating a robotic arm seen via cameras on a local television screen, allowed researchers to explore the psychological aspects of VR without having access to VR hardware. Steuer further proposed that the two primary factors leading to the experience of presence in another place were "vividness" and "interactivity." Subsequent research by many others would build on this, defining "immersion" and "interactivity" as two primary components of "presence" (Witmer & Singer 1998). While Steuer's definition of virtual reality was helpful for researchers for over two decades, a new wave of "goggles 'n' gloves" virtual reality headset suggests that we might focus some attention specifically on them.

Virtual reality prototypes have been developed and studied by researchers for decades. One of the earliest examples, developed in 1985 by Jim Humphries and Mike McGreevy, was developed at NASA Ames and sits in the National Air and Space Museum. Viable head-mounted displays have only recently become available to the public. Early attempts in the 1990s to sell virtual reality gaming systems (i.e., the Nintendo Virtual Boy or the Sega Genesis VR) were unsuccessful. In 2013, the success of the Oculus Rift Developer Kit 1 renewed public interest in virtual reality (Figure 4.6). Several technology hurdles made this viable where previous attempts had failed, including screen resolution, refresh rate, and field-of-view. A year later, Google released Google Cardboard, a new developer platform in which smartphones could be used for virtual reality. By 2016, Google began offering an improved version of this (Google Daydream) and three PC-tethered HMDs were available: the Oculus Rift CV (commercial version), the HTC Vive, and the Sony PlayStation VR headsets.

By 2018, the prices for these dropped to make them more accessible to the public: with the Oculus Rift and HTC Vive selling for under $500 and the Sony headset selling for $200 (but requiring a $350 PlayStation 4). Many companies have recently launched or are planning stand-alone VR headsets that don't require a smartphone or PC, including an improved Google Daydream, the Oculus Go (~$200), the Lenova Mirage Solo (~$400; running Daydream), the HTC Vive Focus (~$600), and the Oculus Quest (~$400).

4.3.2 How Can Virtual Reality Be Used for Astronomy Education?

A 2007 article entitled "A Desktop Virtual Reality Earth Motion System in Astronomy Education" (Chen et al. 2007) focused on what they termed "Desktop

Figure 4.6. The author with the Oculus Developer Kit 1 prototype.

VR." They defined "virtual reality" as "a real-time graphical simulation in which the user interacts with the system via analog control, within a spatial frame of reference and with user control of the viewpoint's motion and view direction," a definition they adopted from Moshell & Hughes (2002).

While many examples of "Desktop VR" have existed for years (Chen et al. 2007), we are only just beginning to see examples of "goggles 'n' gloves" being used in virtual reality. The most obvious reason to include virtual reality in an astronomy education is telepresence. New VR platforms market themselves as offering the ability to be a virtual astronaut, to view the surfaces of other planets, or to fly through the solar system. In 2014, one of my own graduate students created a prototype program, Luna, for the Oculus DK2+Leap Motion system that placed users high above the Earth's north pole and showed the Moon moving through its phases around them. It then shifted the perspective to the Earth's surface. Switching from an *exocentric* to an *egocentric* view the system can help teach such geometries better than simply using one or the other (Salzman et al. 1999).

One key advantage of VR for astronomy education is the ability to truly see things in three dimensions and from multiple perspectives. As Yu (2005) noted when advocating for full-dome videos in planetariums, "[t]raditional astronomy teaching is made even more difficult by the fact that much classroom instruction involves 2D pictures, charts, slides, and written descriptions in textbooks. For instance, most of the past research on teaching phases of the Moon have used 2D drawings and diagrams. It is usually up to the student to conceptualize 3D abstractions using 2D descriptions." Pre- and post-testing of students in immersive planetariums has shown some improvements in astronomy learning gains over non-visual or even flat-screen visuals (Yu et al. 2017).

Shortly after the debut of the Oculus CV, several smaller companies offered low-cost "tour of the solar system" VR programs, such as Titans of Space 2.0 ($8, DrashVR), Spacetours VR ($7, Vibrant Visuals), and Discovering Space ($10, DiscoveringVR) (Figure 4.7). The proliferation of these stems partially from their ease in creation. The virtual environments are simple: giant spheres with overlaid textures and an interface for navigating from planet to planet. More unique astronomy- and space-related options include Apollo 11 VR (Immersive VR) and Mission:ISS (Oculus Studios) which both allow users to follow footsteps of the real space explorers. Virtual reality apps such as Star Chart (Escapist Games) can also allow users to identify constellations in Earth's sky.

The state of these new VR experiences mirrors the state of desktop planetariums in 1992 when programs like The Sky and Starry Night emerged. Those still exist, but today the open-source program Stellarium dominates the market. Likewise, one might expect that in the future something similar, but better, will emerge to become the dominant VR space exploration program. The open-source WorldWide Telescope, created by Microsoft and now managed by the American Astronomical Society, offers some VR functionality, though its emphasis is more on star show and video creation and less on real-time user exploration.

Figure 4.7. A guided VR tour of the solar system in Discovering Space 2. ©DiscoveringVR: Discovering Space 2.

Many of the astronomy-related VR experiences listed already are passive or include a minimal degree of interaction. Some of these applications purposely limit user motion to accommodate VR users that don't have head-tracking (e.g., the Samsung Gear). Others applications that are intended for the latest HMDs (e.g., Mission ISS) allow interactivity such as moving through the scene or moving objects. One program that allows for user creativity is Universe Sandbox 2. This program is a VR port of an earlier game designed for simulating the collision of galaxies.

While not intended specifically for astronomy education, there are a handful of 3D-drawing programs for VR, most notably Quill, Medium (both by Oculus), Blocks, and Tilt Brush (both by Google). Tilt Brush, in particular, is well-suited for easily allowing students to create astronomy-related objects (Figure 4.8). The brush palette includes light, fire, clouds, and stars which are all perfectly suited for drawing galaxies. It also included special effects which allowed easy representation of particles and their motion.

In my "Astronomy 101" class, I had student teams of three work in roles: script writer, video producer, and VR painter to make videos that introduced an astronomy topic and then showed their 3D painting of it. The majority of the 16 teams chose Google Tilt Brush for the reasons stated above. Five teams chose to use Oculus Medium; one team chose to construct a model of the Huygens probe to Titan using Google Blocks. While tools exist for easily creating 3D objects using VR, in 2018 there is not yet a simple way to work collaboratively in groups *within* VR or to share these builds with a broader, real-time audience as was possible a decade ago with Second Life.

4.3.3 Getting Started in Virtual Reality

To understand the potential of virtual reality, one really needs to try it first-hand. While low-cost options for virtual reality exist, they have issues that make them problematic for student use. One serious problem with the early VR prototypes and the current low-

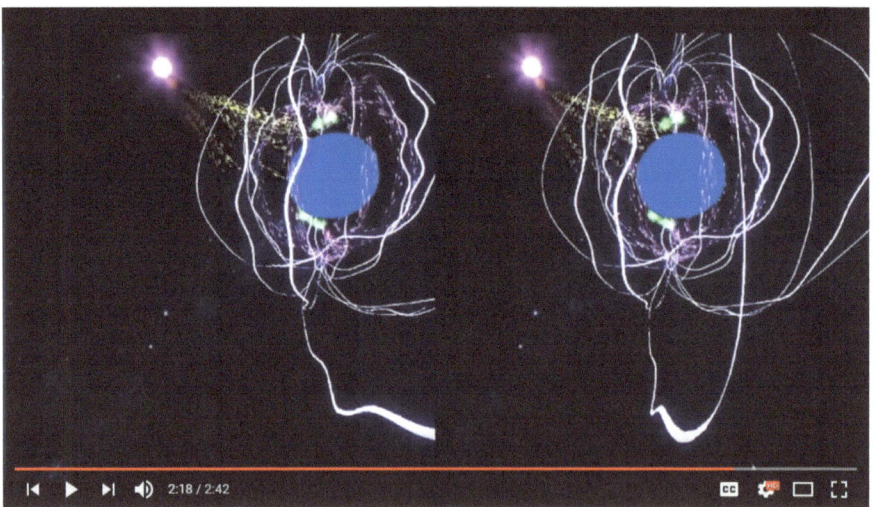

Figure 4.8. A 3D painting of the Earth's magnetosphere created by students using Google Tilt Brush.

cost headsets is the sense of nausea they can induce. People that have a bad experience with VR on the first try are prone to say "I'm not a VR person" when offered a second opportunity. Industry experts worry that low-cost headsets, such as the $10 Google Cardboard effectively "poison the well" by permanently scaring off potential users (Trenholm 2014). The first Oculus Developer Kit (DK1) only had sensors to measure head tilt (φ, θ, ψ) and lacked any external cameras to measure where the user's head was in space (x, y, z). The second kit (DK2) improved on this with one camera that measured head position. In a small study, Tyrrell et al. (2018) found when people used the Oculus DK2, only 10% experienced moderate nausea. By comparison, almost three times that many experienced the same degree of simulator sickness when no external camera was used (Treleaven et al. 2015). The current Oculus CV supports up to four cameras to track the user, reducing this problem even more.

If, as educators, we don't want to "poison the well" we should only use multi-camera VR hardware. This likely means purchasing a single HTC Vive or an Oculus CV with an extra camera rather than 10 cheaper units (such as the Oculus Go for $200 each) or 200 headsets that require separate smartphones (for $10 each). Currently, most of the cost of a room-scale system is for a "gaming PC" that has an adequate memory, enough USB 3.1 ports, and a powerful graphics card. The minimum specifications can be found on the Oculus and HTC Vive website, though obviously buying above spec will improve the performance. In 2018, the cost of a computer that could adequately support VR ranged from $1000 to $2000.

Given that the per station cost of VR hardware, including installation, could be as much as $3000, it is prohibitively expensive to create many VR classrooms. The space suggested per student ($10' \times 10'$ per station) also requires an unusually large classroom. Having a few publically available stations, in someplace such as the library or makerspace, allows instructors to assign VR work as homework rather than treating VR as an in-class activity.

That said, even with the best VR hardware, any VR experiences captured with a video camera (as opposed to those generated with a computer) can still induce nausea due to motion parallax. While Google Lightfield technology may solve this eventually, for now, there is no good way to create a 3D, 360° VR video of some event that doesn't suffer from motion parallax.

Fortunately, some of the best VR tools for creating 3D content are available for free. As described above, Google Tilt Brush, Google Blocks, Oculus Medium, and Oculus Quill all provide interesting ways to create 3D objects. These can be placed into a VR scene using Unity 3D, which creates VR content but doesn't require a VR headset to use. It also is available to students for free. Thus, students and instructors can work on assembling their 3D scenes with Unity 3D even when they don't have access to the VR headset.

4.4 Virtual Reality Virtual Worlds

4.4.1 What Are Virtual Reality Virtual Worlds?

Both the fictional Metaverse and OASIS were *virtual worlds* software that users experienced using *virtual reality* hardware. Only recently have actual virtual reality virtual worlds (VRVWs) become available. Linden Lab briefly released VR-support for Second Life but then withdrew it. Shortly after, both Linden Lab and its former president each offered competing virtual reality successors to Second Life: Linden Lab's Sansar and Philip Rosedale's High Fidelity. The former includes a NASA Apollo Museum and a recreation of the Apollo 11 landing site, both created by the video game company LOOT Interactive (Figure 4.9).

The platform AltspaceVR emerged as a popular early VRVW, declared that it had run out of money and was going out-of-business in 2017 August, and was then acquired and resurrected by Microsoft in 2017 October. AltspaceVR allows for the creation of custom spaces and can be viewed using nearly all existing VR hardware or a 2D display running Windows. Thus, it is easily usable on multiple platforms. Unlike with Second Life, this must be done outside of the program itself using A-Frame, an open-source WebGL language. One of these is a sketched sky of constellations and virtual orrery. AltspaceVR also allows users to create events and space-themed ones appear semi-regularly, including interviews with Bill Nye and viewings of SpaceX launches.

Another cartoonish VRVW is Rec Room created by the Seattle-based company Against Gravity. Rec Room originally required a VR headset. However, in addition to supporting both the Oculus and HTC Vive, it also works with the PlayStation VR headset which has sold more units than Oculus and HTC combined. Rec Room was originally designed as a place for simple kids' games such as paintball or 3D Pictionary. In late 2017, the option to create personalized clubhouses was added to Rec Room. In theory, the Maker Gun tool allows for the creation of objects inworld, though with a very limited color and shape palette (Figure 4.10).

Figure 4.9. Apollo Museum within Sansar virtual reality virtual world.

Figure 4.10. The Maker Tool within Rec Room.

All of the VRVWs mentioned above have failed to attract many users (Table 4.1). At any given moment, somewhere between 5 and 50 people around the world might be using them. By far the most popular VRVW in 2018 is VRChat. During the 2017–2018 holiday season, the number of VRChat players spiked from 100,000 to 3,000,000 in just three months.[1] Unlike with the other VRVWs, both VRChat avatars and environments must be created and customized outside of the program itself and imported using the industry standard game engine, Unity 3D.

4.4.2 Getting Started in Virtual Reality Virtual Worlds

A decade ago, Second Life served as an interesting new platform for universities, museums, and government agencies to explore new methods for formal and informal astronomy education online. The dawn of commercial VR offers many new opportunities for this. As mentioned above, several of the VRVWs already have some astronomy-themed areas or events. No known efforts have been made to use VRVWs in astronomy education yet. Thus, in this section we explore several topics that should be considered in any future projects by educators and developers.

---

[1] Curiously, this was largely driven by a popular and bizarre "Ugandan Knuckles" meme. Knuckles is a character from the Sonic the Hedgehog video game series. A video game review briefly featured a cartoon sketch of short, squat Knuckles. This spawned a series of videos of that cartoon Knuckles singing, which then prompted another user to make a 3D model of the short. This 3D model was combined with audio clips from the low-budget Ugandan action film, "Who Killed Captain Alex?" to make a character in VRChat dubbed "Ugandan Knuckles." Thousands of users would change the avatar to Ugandan Knuckles, circle around new players, and repeatedly ask, "Do you know the way, brother?" Failing to respond led the circle of Knuckles to pretend to spit on the target; leading the Knuckles to someone else dressed as a jar of whey protein (a homonym for "the way"). The Knuckles would eventually move on to another target. Some players would broadcast this trolling on the videocast platform, Twitch, allowing anyone with a smartphone to watch, even if without a computer fast enough to run VRChat. Thus, the number of users of VRChat was able to grow exponentially as this meme spread virally.

Table 4.1. Characteristics of Various Online Learning Environments

	Concurrent Users (April 2018)	Events Calendar	Avatar Appearance	Streaming Video	Non-VR Options
AltspaceVR	~15	Yes	Cartoon	Yes	Windows, Mac
Rec Room	~400	No	Cartoon	Yes[a]	Windows, Playstation[a]
Sansar	~5	Yes	Customizable	Yes	Windows
SineSpace	~10	Yes[b]	Customizable	Yes	Windows, Mac, Linux
VRChat	~5000	No	Customizable[c]	Yes	Windows
Second Life	~40,000	Yes	Customizable	Yes	Windows, Mac, Linux

Notes: [a] This is only available for "registered players in good moderation standing."
[b] User events must be sent to SineSpace admins for posting to the Events Calendar.
[c] Avatars must be created outside of the VRChat and uploaded using Unity.

Age Appropriateness—Not all virtual worlds have been appropriate for all age groups. The extent to which users could shape the environment within Second Life, coupled with the anonymity encouraged when creating an account, led to a heavy emphasis on sex and gambling that was inappropriate for children. In the mid-2000s, the main grid of Second Life was limited to those 18 years and older; younger users were relegated to a "Teen Grid." Virtual worlds targeted at children have drawn larger audiences, presumably since children had free time for such games. For example, Club Penguin reported having almost ten times as many registered accounts as Second Life, with 200 million registered accounts in 2013 versus 21 million accounts for Second Life in 2010. Unless educators know that all of their potential users are over 18, they may choose to develop for non-explicit VRVWs that restrict adult content and monitor in-game behavior. However, if the target audience is adults, then they likely wish to avoid kid-friendly VRVWs (e.g., Rec Room) since too many children can drive away adult users.

Hardware Requirements—One obvious obstacle to experiencing virtual reality is access to the necessary hardware. A decade ago, Second Life and similar VWs did not work well unless users had a computer with a moderately fast graphics card. This limited its user base to those that had the money to afford a nice PC. The low-resolution virtual world of Minecraft, by contrast, became one of the most successful games of all time, partially because it could be played on almost any computer. To increase their user bases, some VRVWs offer a non-VR option so that users can explore the virtual world in a limited way with a standard computer. For example, Sinespace offers desktop options for Windows, Mac, and even Linux. Unless educators can provide ready access to the necessary VR hardware and operating system, they may wish to choose VRVWs that allow for participation with a regular computer and multiple operating systems.

In-World Events—Users of virtual worlds are drawn to them primarily for friendship (Hassouneh & Brengman 2014). With relatively few concurrent users in any given virtual world, it can be difficult to have a critical mass that gives users the sense they are part of a community of learners. The ability to create and advertise in-world astronomy-themed events helps to overcome this, often drawing dozens of users to a virtual room that might otherwise have few to none. Three-dimensional versions of an "Ask Me Anything" (which is popular on Reddit) already take place in AltspaceVR, with celebrities such a Rick and Morty creator, Justin Roiland. Even a simple "watch party" for a stream of astronomy-related video content can draw people. For example, in AltspaceVR which typically has only about 15 users total at any given moment drew 15 people to a viewing of a SpaceX launch. Surprisingly, the current most popular VRVW, VRChat, does not offer the ability to advertise events, seriously limiting its potential for astronomy outreach. While events are not necessary for formal education (since students can be told when and where to go by the instructor), informal astronomy educators can easily use publicized VRVW events as a way to draw an audience.

Avatar Customization and Trolling—Online "trolling" where users attempt to annoy other users is common on the web, including in virtual worlds. Public events in Second Life were sometimes trolled, most notably when a series of flying penises disrupted a C-NET interview with virtual world celebrity, Anshe Chung (Au 2016). Some of the trolling was verbal; some of it was visual. Muting other users is relatively easy and common in most VRVWs. However, virtual worlds with the option for users to completely customize their avatars (as VRChat does) easily allow trolling with visuals. Most VRVWs in 2018 offer limited customization of cartoon-like avatars (i.e., AltspaceVR, Rec Room). Educators wishing to minimize trolling will want to operate in VRVWs that enforce a digital "dress code" and do not allow completely customized avatars.

World Customization—Some VRVWs, including VRChat and Sansar, allow people to easily upload virtual experiences using common 3D development tools such as Unity 3D. Rec Room offers limited room customizations using tools within the application. AltspaceVR allows users to create customized environments using the WebVR language, A-Frame.

Student Issues—As mentioned earlier, a non-negligible fraction of VR users have experienced nausea with early VR prototypes and cheap, cardboard headsets. These users are naturally reluctant to try VR again. In my astronomy classes with 48 students, one was uncomfortable having something attached to his face. In that same group, another student spent hours sculpting in VR but then was very anxious about walking around her sculpture, in spite of the face that she was in an empty room with "spotters" there for her. One student enjoyed using the headset but sweated enough that it needed time to dry out rather than just being cleaned. Many students will experience extra discomfort with VR headsets if they wear glasses. These are unchartered waters for both instructors and students, it may not be advisable to *require* all students to wear a VR headset.

The Future Cost of VR—Currently, both the user accounts and the ability to create experiences in all of the VRVWs mentioned is free. Each of them is attempting to draw in customers and stake out a place in the market. Eventually, the companies will likely need to monetize these. Second Life earns its money from selling virtual land; astronomy educators needed to lease this virtual land (actually server space) in order to build the virtual educational spaces. This is a non-negligible cost (~$300 per month) and might be difficult to justify given the low number of users. When the current generation of VRVWs begins charging, institutions will need to consider the cost-benefit ratio. For now, most VRVWs are run by relatively small companies: only one VRVWs is owned by a large company: Microsoft's AltspaceVR.

4.5 Concluding Remarks

As we enter an era of where VR is readily available in higher education, it is helpful to reflect on lessons learned from virtual worlds, from serious games, and from astronomy education, in general. **What kind of virtual reality experiences should we encourage to enhance and promote astronomy education in both formal and informal settings?** While the VR technology is quite new, the research and experience of educators in virtual worlds certainly has much to tell us about how students might respond. The examples listed above, both for VWs and VR, give clues to what might be possible in the decade ahead. That said, it is equally easy to use VR to perpetuate ineffective educational techniques; the technology doesn't make it magically more impactful. In crafting new experiences for students, educators and game designers should keep the repeating mantra from Slater & Adams (2003) in mind: **It's not what the instructor does that matters; it's what the students do.**

References

Au, J. W. 2016, What Dealing With Second Life Griefing Can Teach Developers Of New VR Social Worlds [Blog post], New World Notes, http://nwn.blogs.com/nwn/2016/11/facebook-jim-purbrick-linden-lab-vr.html.
Bary, J., & Frank, A. 2018, At Play in the Cosmos: The Videogame (New York: W. W. Norton)
Bell, M. W. 2008, J Virtual Worlds Res, 1, 2
Bloom, B. S. 1956, Taxonomy of Educational Objectives, Handbook I: The Cognitive Domain (New York: David McKay Co)
Chen, C. H., Yang, J. C., Shen, S., & Jeng, M. C. 2007, J Educ Technol Soc, 10, 289
Conklin, M. S. 2007, 101 Uses for Second Life in the College Classroom, https://learninginnovation.duke.edu/pdf/tools/sluses.pdf
Crider, A., & Torrez-Riley, J. 2017, J Virtual Worlds Res, 10, 7
Crider, A. W. 2006, BAAS, 38, 1032
Duncan, S. C. 2011, Well Played: J Video Games Value Mean, 1, 1
Finkelstein, N. D., Adams, W. K., Keller, C. J., et al. 2005, PRSTP, 1, 010103
Gauthier, A. J., & Impey, C. 2008, BAAS, 40, 236
Green, P. J. 2003, Peer Instruction for Astronomy (Upper Saddle River, NJ: Pearson Education)
Hassouneh, D., & Brengman, M. 2014, Comput Hum Behav, 33, 330
Jarmon, L., Traphagan, T., Mayrath, M., & Trivedi, A. 2009, Comput Educ, 53, 169
Krathwohl, D. R. 2002, Theory Pract, 41, 212
Laughlin, D. 2008, Request for Proposals to Partner with NASA on the Development of a Massive Multiplayer Online Game to Support STEM Learning, https://prod.nais.nasa.gov/eps/eps_data/129682-OTHER-001-001.pdf

Moshell, J. M., & Hughes, C. E. 2002, in Handbook of Virtual Environments: Design, Implementation, and Applications, ed. K. M. Stanney (Mahwah, NJ: Lawrence Erlbaum), 893
Salzman, M. C., Dede, C., Loftin, R. B., & Chen, J. 1999, Presence, 8, 293
Schiller, S. Z. 2009, J Inf Syst Educ, 20, 377
Schroeder, R. 2008, J Virtual Worlds Res, 1, 2
Slater, T. F., & Adams, J. P. 2003, Learner-centered Astronomy Teaching: Strategies for ASTRO 101 (Boston, MA: Addison-Wesley)
Steuer, J. 1992, JCom, 42, 73
Tateru, N. 2012, No More Quarterly or Annual Second Life Economy Reports, says Lab [Blog post], http://dwellonit.taterunino.net/2012/03/01/no-more-quarterly-or-annual-second-life-economy-reports-says-lab/
Treleaven, J., Battershill, J., Cole, D., et al. 2015, Virtual Real, 19, 267
Trenholm, R. 2014, Oculus Warns Rivals Not to 'Poison the Well' with VR Kit That Makes Us Sick, https://www.cnet.com/news/oculus-warns-rivals-not-to-poison-the-well-with-vr-kit-that-makes-us-sick/
Tyrrell, R., Sarig-Bahat, H., Williams, K., Williams, G., & Treleaven, J. 2018, Virtual Real, 22, 211
Winkelmann, K., Keeney-Kennicutt, W., Fowler, D., & Macik, M. 2017, JChEd, 94, 849
Witmer, B. G., & Singer, M. J. 1998, Presence, 7, 225
Yee, N. 2006a, CyberPsychol Behav, 9, 772
Yee, N. 2006b, Presence, 15, 309
Yee, N. 2017, Beyond 50/50: Breaking Down The Percentage of Female Gamers by Genre [Blog post], https://quanticfoundry.com/2017/01/19/female-gamers-by-genre/
Yu, K. C. 2005, Planetarian, 34, 6
Yu, K. C., Saham, K., Sahami, V., Sessions, L., & Denn, G. 2017, in 2017 IEEE Virtual Reality (VR), ed. E. S. Rosenberg, D. M. Krum, Z. Wartell, et al. (Piscataway, NJ: IEEE), 237–8

Chapter 5

Massive Open Online Astronomy Courses

Paul Francis

Massive open online courses (MOOCs) are controversial free online courses, designed to be taken by tens of thousands of students. I discuss why you might want to teach a MOOC, and share practical tips and tricks that I have learned in developing and teaching my four MOOCs to over 300,000 students. I provide an overview of the MOOC scene, how astronomy fits into it, and show data on the types of students who study MOOCs, and the very small fraction who actually complete a course.

Chapter Objectives

By the end of the chapter, readers will be able to:
- Understand the types of students who take astronomy MOOCs, and how they progress through a course.
- Explain the reasons why they might want to teach a MOOC.
- Understand best pedagogical practice for developing and teaching a MOOC.
- Find resources for developing and teaching a MOOC.

5.1 Introduction

Massive open online courses (MOOCs) are a controversial topic at most universities. Some see them as a disruptive wave of the future, posed to do to universities what Uber is doing to taxis. Other point out that online education long pre-dates MOOCs, which they regard as an attempt to hype something that wasn't really new.

How does a MOOC differ from any other online course? Their proponents claim two main points of difference. First, a MOOC must be scalable: capable of being taken by hundreds of thousands of students. Second, a MOOC must be open: freely available to a wide range of students, and not restricted to students at a particular location, or those enrolled in a particular fee-collecting institution.

MOOCs are relatively new, having only come to widespread attention around 2012. The earliest MOOCs gained lots of attention firstly because of the massive numbers of students who enrolled—typically in the hundreds of thousands, and because they came from prestigious high-profile universities, such as Stanford and MIT (e.g., Liyanagunawardena et al. 2013; Ebben & Murphy 2014). There was a mad rush as other universities, worried of missing out, scrambled to launch their own MOOCs.

Some universities have launched MOOCs using their own technology platforms or normal learning management systems such as Blackboard (www.blackboard.com/index.html) or Moodle (https://moodle.com; e.g., https://mooc.utas.edu.au/courses). But most have chosen to join a consortium, using a common technology platform. Consortia such as Coursera (https://www.coursera.org/), edX (https://www.edx.org/), Udemy (https://www.udemy.com/) and FutureLearn (https://www.futurelearn.com/) have come to dominate the MOOC scene, providing well developed technology platforms, and wide visibility to courses.

As of early 2018, over nine thousand MOOC courses have been offered, from over 800 universities, and reaching over eighty million students around the world (https://www.class-central.com/report/moocs-stats-and-trends-2017/). The hype has mostly dissipated, but all the consortia are still hard at work developing their technology platforms, and new courses are released every week.

Astronomers have been enthusiastic adopters of MOOCs, much as they have been of other outreach and communication technologies. There are currently over 30 MOOCs in astronomy related topics. They range in level from very basic introductory courses aimed at the general public, to fairly technical courses aimed at graduate students, though the introductory courses predominate. Exoplanets and alien life are the most popular topics.

To see many of these courses, go to the main edX consortia websites (https://www.edx.org/, https://www.coursera.org/, https://www.udemy.com/, https://www.futurelearn.com/) and search for "astronomy." You can enroll in almost all for free and have a look at how they are set up.

This chapter is intended to be a practical guide for astronomers who are considering developing a MOOC. I'll outline some of the many lessons I've learned in producing my own series of four MOOCs (https://www.edx.org/xseries/astrophysics), as well as lessons gleaned from the literature and from conversations with other MOOC developers.

My four MOOCs were launched in 2014, co-taught by myself and Brian Schmidt, and published through the edX consortium, which our university had joined. The plan was to convert my on-campus core 1st year astrophysics course into a MOOC. The course was intended for students with a strong maths and physics background from high-school. We rapidly decided that a full semester-long course would be indigestibly large for MOOC students, so we broke it up into four independent pieces:

1. Greatest Unsolved Mysteries of the Universe;
2. Exoplanets;
3. The Violent Universe;
4. Cosmology.

The courses have all now been run several times, both in instructor-paced mode and self-paced mode, and have now had a combined cumulative enrollment of over

300,000 students. They have been updated several times, both to improve the pedagogy and to update the astrophysics, and we plan to continue offering them for at least several more years.

5.2 MOOC Statistics

The first wave of MOOCs attracted enrollments of hundreds of thousands of students, but as the number of MOOCs has increased, typical enrollments have dropped and are now typically only in the thousands, though the dispersion in this number is enormous. The current version of my first course (Greatest Unsolved Mysteries of the Universe) currently has 26,000 students enrolled.

These students come from a wide range of countries: over 178 in the first run of this course. Top five countries currently in this course (Table 5.1):

The following statistics refer to the first, instructor-paced run of the "Greatest Unsolved Mysteries" course in 2014, though very similar statistics are obtained for the other courses and other runs of this course.

The students are typically well educated already: 23% have an advanced degree and 35% have a college degree.

Most are older than typical on-campus undergraduates though the modal student is in their twenties (Figure 5.1). The course has been successfully completed by students as young as age 11 and as old as age 92.

Most of the students who enroll in a MOOC do not get very far into it. Around 70% of those who enroll never actually log in—they are presumably using the enrollment process as a way to bookmark a course to come back to it later, but never actually end up doing so. This is particularly prevalent with 20–30 year-olds (Figure 5.2).

Even if a student started my course, most only watched the first 1–3 videos and did not attempt any assessment (Figure 5.3). This is typical (Breslow et al. 2013).

Once a student had completed the first homework assignment, however, they had a 75% chance of going on to pass the course, which is somewhat higher than is typical (Ho et al. 2015). This is comparable to the completion rate of more traditional online courses (e.g., Herbert 2006). Due to the very rapid decline in student numbers during the first few minutes of involvement in the course, however, this type of statistic will probably be highly sensitive to the exact nature and placement of the first homework assignment.

Table 5.1. Countries Represented by Students in This Course

Country	Percentage of Students (%)
USA	27
India	10
United Kingdom	6
Australia	5
Canada	5

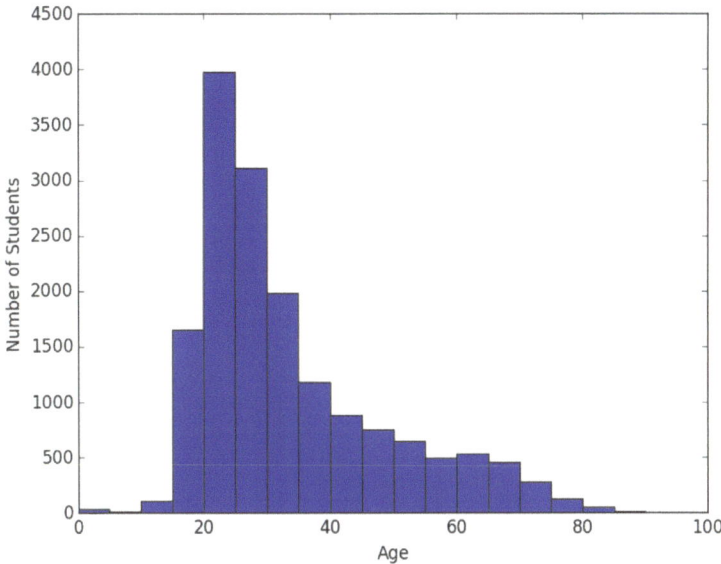

Figure 5.1. Age distribution of people enrolled in the course. Note that the ages are based on self-reported birth dates (Francis 2015).

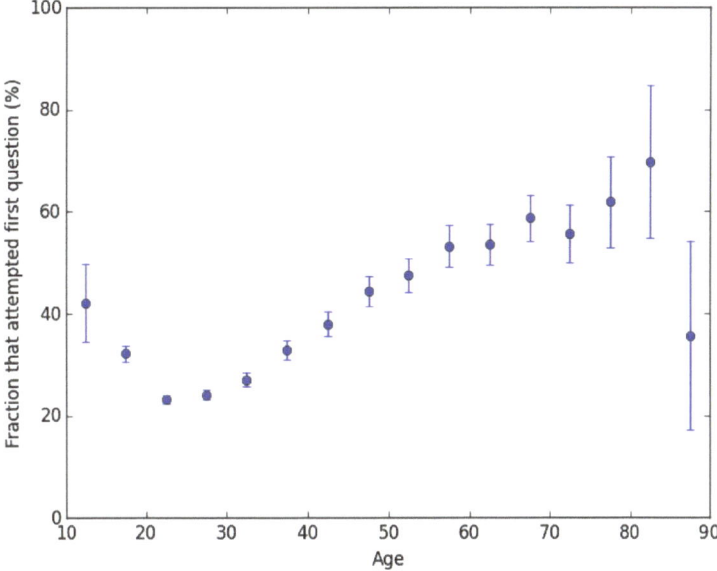

Figure 5.2. Probability of an enrolled student starting the course, as a function of age, as measured at the start of its first run (Francis 2015).

Most MOOCs show very similar patterns (e.g., Onah et al. 2014), though systematic studies of large numbers of courses have been hamstrung by student privacy requirements. Jordan (2015) showed that the completion rate is highest for short courses, for courses with purely machine-graded assessment, and in more recent courses.

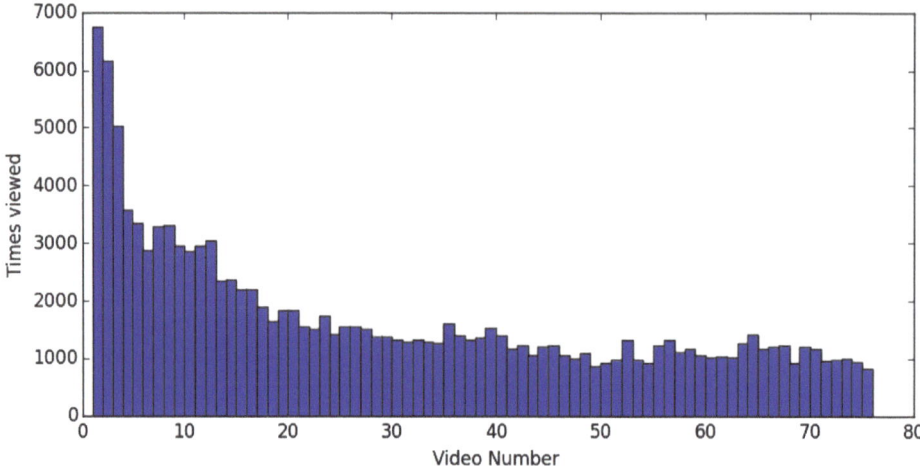

Figure 5.3. Number of times each video lesson in the first course was viewed (measured as the end of its first run) (Francis 2016).

It is thus clear that traditional completion rates (number of passing students divided by the number of enrolling students) are very low, but that this statistic does not mean the same thing in MOOCs as it does in on-campus courses. Enrolling in a course costs a student nothing, and many students just do it as a way of bookmarking a course for future reference (Ho et al. 2015). Even students who start the course do so with many different intentions: at the start of my first course, 44% of the students said they had no intention of doing any assessment. Once the initial rush is out of the way, however, we are left with a core of students who have a very respectable completion rate.

For more details on all these statistics, see Francis (2015).

5.3 Why Teach a MOOC?

MOOCs are usually free, or at most charge a token amount for some certificate. So they do not directly make money for universities. Why then would any university invest time and effort in developing MOOCs? In practice, there seem to be four main reasons:

1. Publicity. MOOCs reach huge numbers of students in many parts of the world, and so MOOCs may be a good way influence people who are otherwise hard for your university to communicate with. 23 million new learners signed up for their first MOOC in 2017 (https://www.class-central.com/report/moocs-stats-and-trends-2017/). I surveyed the students in my courses, and over 50% had never heard of my university before starting the course.

2. Recruitment. I cross-correlated the list of students who took the MOOCs with those who arrived the following year to study our on-campus courses, and surveyed these students to make sure that the MOOCs were the reason why they chose to come to our university. We found that roughly one in

every thousand students enrolling in the MOOCs subsequently enrolled in an on-campus course, with this number split roughly equally between international and domestic students. This will be a substantial underestimate of the true recruitment rate, as some students will arrive after a delay, or will be advised to consider our university by a friend or relative who did the courses.
3. Technology Development. Involvement in MOOCs is a chance for a university to build expertize and capacity in technology-rich online education. The techniques learned can be applied to on-campus courses. In our case, we used what we'd learned from developing the MOOCs to develop a flipped classroom pedagogy for our on-campus physics courses, which has been very successful. Developing some expertize in online education is also a risk mitigation strategy, in case MOOCs really do start to disrupt the university sector at some point in the future.
4. Philanthropy. MOOCs are a way to provide high quality university courses to many people who would not otherwise be able to access them. For example, poor students in developing countries, students who are too young or two old to go to a traditional university, students in remote areas. Many of the students in my courses lived in places with no informal astronomy education and seemingly nobody else who shared their passion for astronomy—taking part in the MOOCs allowed them to join a worldwide community. Some quotes from students:

"I had been suffering from severe depression, crippling [...] I suffered a horrible environment that discouraged any sort of creativity and critical thinking. Your course got me out of that rut and eased me into learning again. To say my life changed by taking this course might be an understatement." (Indian MOOC student, who is now about to start a MSc in astrophysics.)

"Completing these MOOCs had been about returning to a childhood passion for astronomy; or overcoming our fear and insecurity in the face of mathematics; or proving that yes, we can do rocket science; or, post-retirement, finding an activity full of meaning." (Group of Australian MOOC students.)

In recent years, MOOC providers have been experimenting with different ways to raise money from the courses. One approach is to offer a free course as a "loss-leader" for a paid online course. The first fraction of some course or program is offered for free, and if you pass that you can count it as partial credit toward a paid credential. This approach has been used by Deakin University in Australia and Arizona State University in the USA. Another approach is to offer a credentialed course or program, aimed at vocational skills for continuing learning. So far, most of these are "micro-credentials," but the first MOOC degrees are starting to appear.

For you as an individual educator, MOOCs are a chance to reach a wider audience. The number of students who have completed my MOOCs is ten times larger than the number of students I've taught face-to-face during my entire career (and I teach large face-to-face courses). The students are also typically a pleasure to teach—interested, motivated, inquisitive and usually very capable. The unmotivated students would never sign up for a MOOC, or if they did, they would be unlikely to get very far into it. And the top-end students in the MOOCs are academically amazing—even the best students in my on-campus courses wouldn't rank in the top 50 of the MOOC students. That is the benefit of drawing your students from a larger pool.

Astronomy is unusual among the sciences in having a really strong public outreach effort. How do MOOCs fit into that? The MOOC enrollments are large compared on on-campus courses, but still pretty small compared to the numbers of people who might access a NASA web page, watch a space documentary or see a new snippet about the latest astronomical discovery on the TV news. I see all astronomy outreach and education fitting on a spectrum of impact against numbers. At one extreme are PhD programs, where we have a very large impact over many years on a very small number of students. Then you have undergraduate classes (more students, less impact), intensive forms of outreach such as evening classes and public talks, then various forms of media and online outreach, culminating at the other extreme with news clips in the TV news (huge number of people, but very limited impact). MOOCs fit into this continuum somewhere between on-campus courses and media work. They allow you to have a big impact, and to present a coherent body of information at a much greater depth than is possible in the media. But they place considerable demands on the students, which means they will never reach the numbers that more entertainment-oriented resources can.

5.4 Which MOOC Platform?

If you wish to produce a MOOC, you will need to choose which software platform to use, and get access to web-servers with enough capacity to handle the number of students. If you are lucky, your university will already be a member of one of the big MOOC consortia, such as edX, Coursera or FutureLearn, or can be persuaded to join one of these. In this case, the consortium will provide the technology platform, host the servers and give training and support.

But what can you do it your university is not in one of these consortia? Most consortia only allow courses contributed by faculty of member institutions, the exception being Udemy, which allows anyone to contribute a course (https://www.udemy.com/teaching/).

Another possibility is to use a conventional learning management system, including commercial products like (https://www.blackboard.com/index.html) or Google Classroom (https://classroom.google.com/) and open-source products like Moodle (https://moodle.com/) or Sakai (https://sakaiproject.org/). These are not designed to handle hundreds of thousands of students, but the active enrollment in most MOOCs these days will only be in the hundreds or thousands, and can be feasibly handled by normal learning management systems. Most educational institutions today will already have access to at least one of these.

A third possibility is to use open-source software. The EdX consortium release their software freely (https://open.edx.org/). You can run it on your own servers, or there are commercial providers who will run it for you. Open-source learning management systems such as Moodle (https://moodle.com/) are another option, though they won't be able to handle very large class sizes.

5.5 What Topic Should I Teach?

OK—let's say you've decided to teach a MOOC. Which topic should you cover? What level should your course be pitched at? And how long should the course be?

This will depend crucially on your motivation for teaching the MOOC. If your primary motivation is publicity or recruitment, then you should teach an introductory-level course, as that will reach the maximum audience. The course should be short, because that means more students will be prepared to take it, and it also minimizes your workload in developing the course.

If, on the other hand, your motivation is philanthropy, then you may wish to pick a topic that does not overlap with existing MOOCs. At present, exoplanets and life in space are the most popular MOOC topics, so an additional course in this area would have to be significantly different or better.

Most existing astronomy MOOCs are at the "ASTRO101" level (1st year, introductory, minimal maths and physics pre-requisites). There may be a case for developing more MOOCs at a higher level. They will not reach such a wide audience, and so are less useful for publicity purposes. But they can perhaps be used for other purposes—such as exchanging them with other universities. The biggest teaching load at many universities is all the specialist later year and graduate courses —wouldn't it make sense to have each university produce a really good MOOC in their area of special expertize, and share it with other universities? They could be used for flipped classes, in which students get their content from the MOOC but then work in-person on more complex problems, during group tutorials or workshops.

Another argument for teaching higher-level courses is so as not to compete with the vast quantity of outreach materials available. An introductory course is competing with TV documentaries, many websites, and many popular books. But for keen students who have mastered the popular level material, there aren't many places to take their knowledge of astronomy to the next level. MOOCs could provide that.

How long should your course be? A typical university course in Australia runs for roughly 12 weeks and requires roughly 10 hr of student intellectual activity per week. This is typically made up of some combination of lectures, laboratories, tutorials and homework—i.e., around 120 hr of student work to finish. Some MOOCs are this large, but increasingly they are much smaller—requiring perhaps only 2–3 hr of work per week over maybe 4–5 weeks (~15 hr total?). Short courses have higher completion rates (Jordan 2015). There is no hard evidence for how the course length affects enrollments, but anecdotally many students are reluctant to take on a large long-term time commitment.

On the other hand, if your MOOC is the length of a full course (~120 hr total), you can re-use the material in many other contexts. You could use it with your on-campus students, and exchange it with other universities, and possibly award university credit for it. But the length will scare away more casual students. My MOOCs are offered to the usual MOOC market of people around the world, but we also run them "for credit" as advanced placement activities for school-children, as an accredited course for on-campus students, and via an exchange agreement with students from other universities in the edX partnership.

A common compromise is to break a full university course into sections and turn each section into a separate MOOC. Casual students won't be scared off by the huge time commitment needed to complete the whole course, but keen students can work their way through a series of courses.

5.6 Course Structure

Many MOOCs today share a fairly similar structure. They are divided into sections, which the student takes in order. Within each section, there will be:
- A lesson. This is the equivalent of a lecture in an on-campus course. Typically consists of a series of short (5–10 min) videos, interspersed with questions.
- Assessment. Some questions based on the lesson in this section. May be multiple choice questions, numerical entry questions, peer-marked essays.
- Discussion. A bulletin-board tab to discuss that material in this section with other students.

This structure is basically an online equivalent of the "Lesson, Homework, Tutorial" structure of traditional on-campus classes, and may not be optimal. But it's what most MOOCs use for the moment.

Within this structure, however, there is one key choice to make:

Self-paced or instructor-paced? In an instructor-paced course, sections of the course are released to students in a sequence, and the homework has deadlines. The course has a definite start- and end-date. This means that students proceed through the course together. The main benefit of this approach is that it increases discussion between students, as they will all be looking at the same part of the course at the same time. It also allows a wider range of assessment tasks. And it gives instructors a break between runs of a course.

In a self-paced course, students can proceed through the course at their own pace. This is much more flexible, but limits the types of assessment you can set, and makes for much less lively student discussion.

I've run my courses in both modes, and roughly 50% of the students chose each mode.

5.7 Making the Videos

How much video footage will you need? My experience is that you can cover material far more quickly in videos compared to face-to-face lectures—I typically cover the content of a 50 min lecture in around 20 min of video footage, and I've heard the same from many other MOOC developers. I guess we all spend more time than we think in lectures on crowd control and repeating ourselves.

Guo et al. (2014) have analysed the optimum length of the videos. They find that if videos are much longer than about 6 min, students don't watch them all the way through, or even choose not to start the videos at all. So the recommendation is that videos be short, followed by some other activity like a question or a simulation.

There are many different styles of video used:
- Talking head—a presenter talking to the camera, or a recording of a lecture.
- PowerPoint/code screen recording.
- Tablet capture—Khan-academy style videos where the instructor hand-writes or annotates slides while talking (Figure 5.4).
- Green-screen—where an image of the instructor is placed in front of some graphics (Figure 5.5).
- On-site filming, perhaps at an observatory, or in a lab (Figure 5.6).
- Multi-presenter filming, such as interviews or debates (Figure 5.7).

Guo et al.'s study suggests that the production quality is not crucial. What seems most important is that the videos seem to be personal. Talking head videos work best, for example, if you appear to be talking directly to the students, not lecturing to

Figure 5.4. The author recording a tablet-capture video. I use a Wacom Cintiq tablet connected as an external monitor to my laptop, and a Sennheiser PC8 USB gaming headset. I draw using the Sketchbook Pro software, and record and edit using the Camtasia software. For an example of a tablet-captured video, see https://youtu.be/hFhoezl-75k.

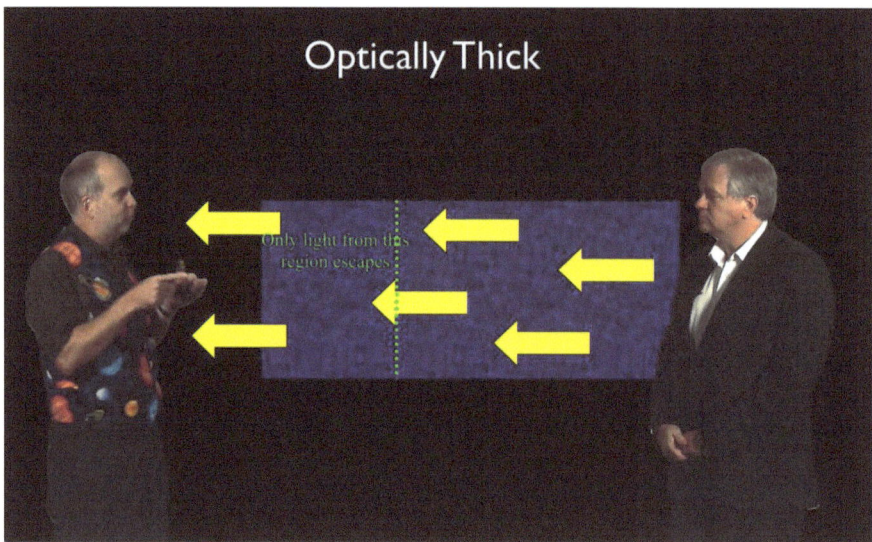

Figure 5.5. The author and his co-lecturer (Brian Schmidt) with a green-screen set-up to insert themselves into their PowerPoint, from the Violent Universe course. These were filmed in a Sitedeck studio owned and run by the university video team. For an example, see https://youtu.be/fiGx6IPka_I. Credit: Paul Francis, ANU.

Figure 5.6. Still from an on-site segment, filmed at Mt Stromlo Observatory, from the Greatest Unsolved Mysteries course. For an example of an on-site segment, see https://youtu.be/3MZKDlMZuJc—it starts with studio filming then moves outside at 1.25 min. This was filmed by the university's video unit using professional equipment, but some segments were filmed by me using a DSLR camera (Canon 640D) and tripod, and a microphone that plugs into my smartphone (RODE smartLAV—http://www.rode.com/microphones/smartlav-plus). Other MOOC teachers have had good results using their smartphone camera, particularly for impromptu interviews. Credit: Paul Francis, ANU.

Figure 5.7. Still from a studio-filmed discussion between the two co-lecturers, and a guest speaker (Brian Cox) from the Cosmology course. These were recorded by the university's video production team using their equipment. An example of a studio-filmed discussion can be found at https://youtu.be/cM3L-c3E_c0. Credit: Paul Francis, ANU.

a hall. And hand-written tablet capture seems more engaging that PowerPoint—perhaps because it encourages lecturers to explain things as if they were scribbling notes on a bar napkin to an individual student, rather than giving a formal lecture.

If you are filming a video yourself, pay particular attention to the sound quality. This is often neglected, but it's actually very easy to get this wrong. For tablet capture, I use a gaming headset, as it allows me to move my head without affecting the volume—for live-action footage, lapel microphones seem to work best.

A crucial consideration here is practicality. If your course is short, you may only be producing an hour or two of video footage. And if your university gives you lot of support with video production and editing, then you can afford to film in a studio. But if you are trying to produce ten or more hours of footage, without much support, then you need an approach you can film in your office whenever you have time, rather than booking in studio sessions. If you want to film outside, you will need good weather and little or no wind—which can be hard to schedule if you need to book a time with a film crew, days in advance.

Another crucial consideration is expectations management. There is a tendency for university administrators to expect you to produce video footage of the quality of (say) a Netflix or BBC documentary, but on a budget only 1% as big!

Producing good videos throws up many surprises. Some people need to work from a detailed script, others can work more free-form. Some people can write on a tablet and talk at the same time, while others can't do this. I'd recommend doing quite a bit of experimentation with different styles early on, before committing to a particular video style. Running full scripted risks giving your videos a wooden quality but running with no script often leads to important points being missed, or to repetition, and you will find you need to film each clip multiple times to get a satisfactory result. Experiment to find what works best for you. For us, it was running semi-scripted: we had a detailed outline of what we would cover but did not read a script. This is another benefit of short video clips—we found we could keep the plan for a five-minute video in our heads and usually film something acceptable in a single take, but longer ones were harder.

In our case, we found that if we ran semi-scripted, little editing was required: just trimming the clips, and maybe joining different short segments together. I did the editing myself, using the Camtasia software. Most MOOC teachers, in contrast, film multiple takes of each video, and require the services of a video editor to combine them into one seamless presentation. If you have a clear plan and script before filming, editing will be minimal, but if you film sequences without a clear idea of how they will fit together, the editing workload can be immense.

I find that worked examples are particularly effective as videos: taking students through problem solutions step-by-step, while talking through your thought processes (Francis 2013).

In our MOOCs, we used two presenters for most of our videos. We did this because my co-lecturer (Brian Schmidt) had a Nobel Prize, but no time, while I had time and educational expertize, but no Nobel Prize. But it turned out to be harder than we had thought—there is a reason why most documentaries have a single presenter. In our early trials we would often repeat material or lose our train of thought. We tried a "Sherlock Holmes and Watson" approach, where one of us would play the expert while the other would ask the dumb questions. But when we tried videos of this out of a test audience, the were found to be cringe-inducingly bad. We tried to find TV shows with two presenters we could emulate, but they are few. Mythbusters was the best example we could find. In the end, we used a semi-scripted approach, where we'd plan who spoke first in each clip, and would talk over what we were going to cover in each clip immediately before filming. We would talk to each other, debate, discuss, and question. Students found this very engaging and commented on how it exposed the thinking processes involved in research. Most MOOCs, by contrast, use a single presenter talking to the camera, more reminiscent of a David Attenborough documentary.

You will need to add closed-captioning to all your videos, for accessibility considerations. This also helps non-native English speakers, as they can run the captioning through a translation program.

5.8 Assessment

Videos get most of the attention, but actually producing the assessment is the larger time commitment in developing a MOOC and is probably the most challenging part pedagogically to get right.

One key distinction to consider is the balance between formative and summative assessment. Formative assessment is intended as a learning experience for the students: online questions lend themselves well to this, as you can use questions with multiple attempts, hints and informative answers to help students learn. However, questions like this are not effective at testing student knowledge and ability in a rigorous way, as students can easily collaborate or use the multiple attempts and hints to guess answers without watching the videos or thinking. I found that when I ran the course for MOOC students, formative assessment worked well. Everyone taking the course is interested, or they would not have signed up. And there is no incentive to cheat. But when I used the same courses with on-campus students or "for credit" students, I had to change the assessment to make it more summative, as I saw evidence of widespread gaming of the system to get marks without work. I restricted the number of attempts allowed, randomized the numerical questions and drew questions randomly from larger pools, to discourage swapping of answers.

MOOC assessment must be scalable—you don't want to be marking 20,000 essays. The most common types are multiple choice questions and numerical entry questions. And this poses a challenge. Unless these questions are very carefully written, they can easily test only low-level comprehension—recall of facts and "plug-and-chug" approaches to mathematical problems.

One possible approach is to use essay-style assessment, but rely on either peer marking or artificial intelligence marking. Peer marking is widely used, but restricts you to instructor-paced courses. It is also time-consuming for the students, and so you can't use it too often. Automated marking techniques (e.g., Valenti et al. 2003) may help with this, but typically require a large training set of human-marked essays.

In the early planning stages for our MOOCs, it became clear than many people across the university believed that an online course had to be inferior to an on-campus course, because it could not stimulate discussion and deep thinking (though there is some evidence that on-campus courses are pretty poor at teaching critical thinking, e.g., Willingham 2007). They felt that the limitations inherent in fully automated assessment meant that the course could only convey facts, and not critical thinking.

We decided to address this by using a simulated mystery universe exercise (Francis 2016). I invented a fantasy universe in each course—a universe very different from our own, but using the same laws of physics. In my first course, for example, I had a universe with bubbles rather than stars in the sky.

In the first section, the universe was presented, using simulated graphics to make it exciting and perplexing (Figure 5.8). The students were urged to discuss what they could deduce from this initial set of data on the course discussion forum, and to then request new data. Each week, as the course ran, I would generate the requested data and post it for the students. The data might include images, spectra, photometry or tables for data. In addition, I would post faked TV interviews and a number of simulated research papers.

At the end of the course, the students were taken step-by-step through the solution to the mystery, by a series of multiple choice, symbolic entry and numerical entry online questions.

The mystery generated a vast number of posts on the discussion forum almost immediately. An extremely lively discussion emerged in each of the four courses, with many students deeply involved in a virtual community, trying to solve the

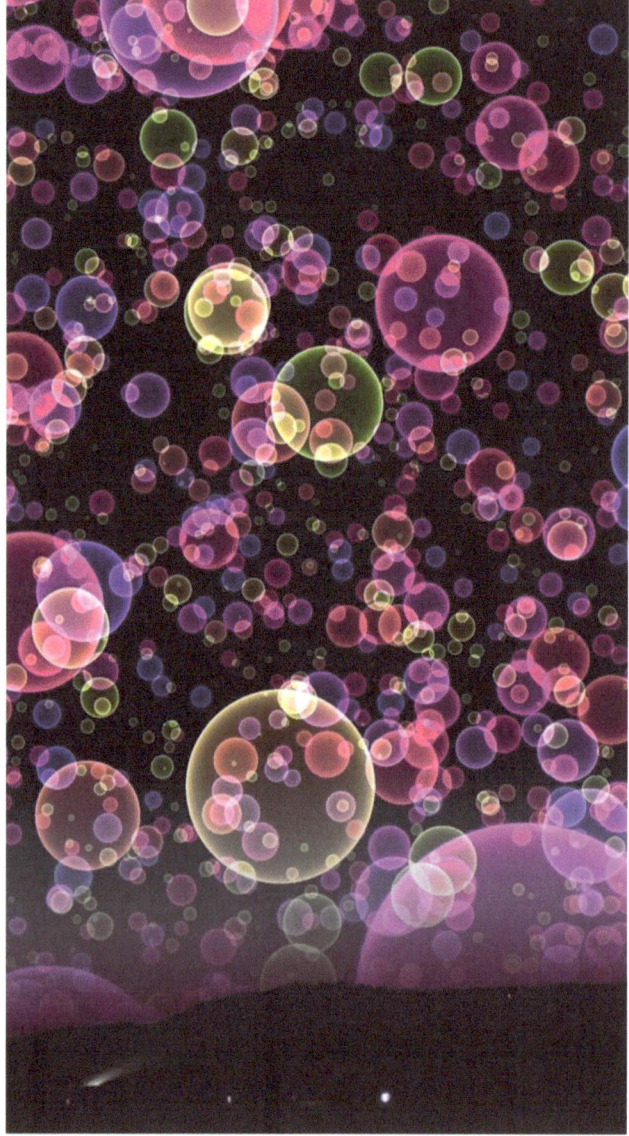

Figure 5.8. Image from the fantasy universe mystery from the Greatest Unsolved Mysteries course—a universe with bubbles rather than stars in the sky. Credit: Paul Francis, ANU.

puzzles. Indeed, the number of posts generated was so large that navigating all the various data requests and theories became a major challenge.

> "Not only did the course present us with the wobbly, complex and frustrating world of research astronomy; it also gave us the chance to feel what this was like for ourselves, through the mystery. If the goal of the mystery was to learn what it was like to be a research astronomer, then high marks shouldn't go to the student

who gets all the answers right first time. High marks should go to a student who looks at the data, then suggests a theory, then gets proved completely wrong, then swears profusely, then devises another set of observations, gets the results, refines the theory, then notices something quite similar in a wholly other branch of astronomy, tries a refinement of the theory on that basis, checks the data again, and ends up with—not the 'right answer', but with at least a model that is close enough to be feeling quite encouraged." (Student quote 2015)

5.9 Discussion Forum

The discussion forum is the main place where you interact with the students, and so it's important to get it right. Ideally you want a lively community using the forum, and answering each other's questions. Some MOOCs achieve this, and others do not. Having an instructor-paced course helps a lot, as does giving rapid responses to early posts and questions. But most of the difference is in the course design.

In our courses (when run in instructor-paced mode), only around 50% of the active students would ever post, and over half the posts were due to a tiny core group of around 10–20 students. However, many more of the students would read the posts relevant to particular questions and topics. Students would normally post for three reasons:

- For help with technical or administrative item (i.e., "when is the deadline for the next homework"). This is most common in the first couple of weeks.
- When stuck on a piece of assessment (i.e., "Any clues how to solve problem 4.03?").
- Discussing the fantasy universe mystery.

Somewhat by chance, our discussion forum worked well. The high-end students would log in to discuss the mystery, and while they were there, they would answer questions from the weaker students, who were typically stuck on some item of assessment.

Some MOOCs make it part of the assessment to post a comment on the discussion forum. I have reservations about this—it boosts the fraction of students who post, but there tend to be a lot of posts with no real discussion, put there purely to meet the assessment criteria.

5.10 Other Things to Think About

One of the biggest challenges in MOOC development was a surprise to me—it's asset management and bug tracking. Our complete set of MOOCs includes over 700 videos, around 2000 questions, and over 1000 pages of text. Keeping track of all of this is tricky, and requires a very systematic approach. You will need some systematic way of naming all these components (e.g., Video V4.03a), and you will need some systematic way to test and track bugs.

We used a two-step testing regime. We first completed one week of the course and invited some trusted colleagues to comment on it—this is the "alpha testing" phase and allowed us to optimize the course structure, mix of filming types etc.

Then, once a course is complete, we paid undergraduate and graduate students to be beta-testers, and run through the whole course watching the videos and trying all the questions. This is where you need a systematic approach to logging all the bugs they find and tasking someone to fix them.

And then, when the course goes live for the first time, the students will find many more bugs, and you will once again have to fix them, now while the course is live, or work around them, while taking systematic note of things to fix once the course run is complete.

The total academic workload involved in producing our courses (equivalent to a full undergraduate course) was about one year full-time equivalent for me, plus about the same time for all the other people involved—testers, film crews etc. Most MOOCs are produced by large teams, and run by a project manager, which can be quite a shock if you are used to the more individualistic way normal university courses run.

You should think early on about your re-use plan. When producing a MOOC, most of the effort is a one-off, up-front workload. Re-using them is very low effort. When we'd completed our first run through our courses, we'd had a cumulative enrollment of around 40,000 students, but now, after running them three times, our cumulative enrollment is over 300,000 students. So plan to re-run your courses lots of time, updating as necessary, and try to get this factored into your budget!

5.11 Conclusions

Is it worth the immense effort needed to develop and teach a MOOC? For myself, I'd have to answer yes. It's been a chance to fashion a wonderful global community of people with a love of astrophysics, and to touch the lives of far more students than I could ever reach in person. It's also been a chance to learn new techniques, which I have applied with great success to my on-campus courses.

Are MOOCs going to take over the world? When I'm optimistic, I note that it is now abundantly clear that for well-motivated students, MOOCs can provide an excellent education at a cost-per-student two orders of magnitudes below traditional courses. In any other field of human endeavor, a product that it a hundred times cheaper while still being as good would take over.

But when I'm pessimistic, I note that there is nothing new about a well-motivated person being able to teach themselves anything. That has been true since the invention of public lending libraries in the 19th century. But how many people ever do this? A strong case can be made that much of university education is about things other than learning (e.g., Caplan 2018), and these other things cannot be reproduced so well by MOOCs.

References

Breslow, L., Pritchard, D. E., DeBoer, J., et al. 2013, Res Pract Assess, 8, 13
Caplan, B. 2018, The Case Against Education (Princeton, NJ: Princeton Univ. Press)

Ebben, M., & Murphy, J. S. 2014, Learn Media Technol, 39, 328

Francis, P. J. 2013, in Proc. The Australian Conf. on Science and Mathematics Education (formerly UniServe Science Conf.), ed. M. Sharma, & A. Yeung (Sydney: Uniserve Science), 136

Francis, P. J. 2015, in Proc. The Australian Conf. on Science and Mathematics Education (2015), ed. M. Sharma, & A. Yeung (Sydney: Uniserve Science), 108

Francis, P. J. 2016, Int J Innov Sci Math Educ, 23, 75

Guo, P. J., Kim,, J., & Rubin, R. 2014, in Proc. First ACM Conf. on Learning@ scale Conf. (New York: ACM), 41

Herbert, M. 2006, Online J Distance Learning Administration, 9, www.westga.edu/~distance/ojdla/winter94/herbert94.htm

Ho, A. D., Chuang, I., & Reich, J. 2015, HarvardX and MITx: Two Years of Open Online Courses (HarvardX Working Paper No. 10)

Jordan, K. 2015, Int Rev Res Open Distrib Learn, 16, 341

Liyanagunawardena, T. R., Adams, A. A., & Williams, S. A. 2013, Int Rev Res Open Distrib Learn, 14, 202

Onah, D. F., Sinclair, J., & Boyatt, R. 2014, in EDULEARN14 Proc., 6th Int. Conf. Education and New Learning Technologies, Dropout rates of massive open online courses: behavioural patterns, 5825

Valenti, S., Neri, F., & Cucchiarelli, A. 2003, J Inf Technol Educ: Res, 2, 319

Willingham, D. T. 2007, Am Educ, 31, 8

AAS | IOP Astronomy

Astronomy Education, Volume 2
Best practices for online learning environments
Chris Impey and Matthew Wenger

Chapter 6

Using New Media and Social Media for Online Learning

Pamela L Gay

New and social media are, for some, the primary ways to consume content and communicate with friends and family. Within each of these entwined technologies are opportunities for the astronomy EPO professional to bring science to the masses and to create pathways to lifelong learning for their learners. That said, there are unique risks to using these media that aren't a concern with other educational techniques; there are issues of privacy, ethical considerations in asking students to use non-school-related platforms, and the risk of content consumption and/or creation expanding to fill all available (and non-available) time. In this chapter, we look at the history of new and social media, and explore ways to engage safely and educate effectively using these technologies with all kinds of learners.

Chapter Objectives

By the end of this chapter the reader will be able to:
- Explain the differences between and origins of new and social media.
- Use strategies for engaging in these technologies effectively without them becoming time sinks.
- Understand ethical considerations in using new and social media.
- Find examples of safe and effective ways to use these technologies in both formal and informal settings.

6.1 Overview and History

In our ever-changing technological world, new and social media are becoming the lingua franca of the youth and of technologically fluent older populations. People of all cultures and educational levels are coming together on various platforms to seek entertainment, education, and community. These audiences' needs are consistent with what astronomy has the potential to provide, and the savvy science communicator can leverage these platforms to reach and retain audiences and to take advantage of teachable moments that may be global in nature.

Within this chapter, new media is used to refer to mass communications tools that take advantage of digital media, such as blogs, pod/vodcasts, and YouTube videos. Like traditional media, these forms of content allow people to subscribe and, while comments on the content may be possible, the primary direction of content is from the content provider to the audience. Social media, on the other hand, is inherently more egalitarian, with each member of a social media platform having the potential to both create content and consume content, and with some platforms requiring people to only form two-way relationships (e.g., being "friends"[1] on Facebook), rather than permitting people to follow someone without them following back (e.g., Twitters model of "followers" and "friends").

New and social media are often lumped together under one title or the other. This mixing is related to their origins. In the late 1990s and early 2000s, numerous platforms popped up that enabled people to self-publish content for free and to easily subscribe to others' content. Platforms such as Friendster (2002) and LiveJournal (1999) both enabled easy content discovery, and also had privacy settings so that contents could be shared publicly or only to friends or sets of friends. Communities built around hobbies, city neighborhoods, and real-world friend groups flourished. These platforms failed to go mainstream, however, because content could only be consumed on the originating platform, and none of them had that one killer feature that could draw in massive populations.

In late 2003 and 2004, the technologies began to come into their own, and new and social media became their own separate platforms. In 2003 August, MySpace launched as a place for people to socially engage in short-form content and to create online profiles to show themselves off on the Internet. A few months later, Facebook launched to Harvard students, with similar goals but a much more constrained layout and function set. These two visually different platforms both found the needed killer feature: they let individuals see into the lives of people they otherwise may not be able to access. For Facebook, this was the ability to check out the life of an otherwise off-limits classmate, while MySpace catered more to celebrities and those who wished to see into their lives. Over time, Facebook added more institutions, and eventually opened to the world. Its structured environment and constant feature upgrades allowed it to dominate, and today MySpace is a largely forgotten relic of social media's beginnings. Facebook, and future social platforms, had come to stay, and quickly became the places where people of all ages gathered to share their lives and peer into the lives of others.

2004 also marked the creation of the Really-Simple Syndication (RSS) protocol. This new way of using xml formatted text to share information made it possible for people to use one piece of software to subscribe to any content on any website that provided an RSS feed. This simultaneously led to the invention of podcasting at Harvard, and the proliferation of blogs (which had previously been using Atom feeds).

---

[1] For the purposes of this chapter, "friend" is used to refer to an Internet relationship based on sharing of content, and does not imply a real-world friendship.

What started as an audio phenomena quickly spread to video, with YouTube launching in 2005. That year also saw the release of iPods that could play video and the inclusion of podcasts in the iTunes store. Quickly, other brands produced their own hardware and new platforms emerged for grabbing content. The new media boom had begun, and anyone could be a producer for the cost of a quality microphone and/or camera.

In the 15 years since this technological boom, the natures of new and social media haven't changed, even as new features and platforms have launched (and sometimes died). Next, we look at how new and social media can be used to communicate about astronomy. On the following pages, you will find practical lessons on how to use new and social media without them using up your day. Ethical considerations are also discussed, as are the research-based best practices that have emerged.

6.2 Getting Started: A Practical Guide

Using new and social media is easy. Using it well is hard. Using it well and growing the audience you desire is hardest of all. Three common failure modes plague educators and science communicators who want to start using new and social media to advance science literacy: (1) people jump in to push content without first consuming content, (2) people allow content consumption and creation to expand to fill all available (and not actually available) time, and (3) people judge their success using inappropriate metrics that don't take into consideration the whimsy of "going viral" and the effect of celebrity.

The problem with jumping in and pushing content is that it isn't possible to know how to communicate to an audience without listening to the audience to learn what that audience wants and needs. Science E/PO professionals are expert communicators who are generally specialized. Museum professionals have different expertize than middle-school classroom teachers. People working with economically at-risk populations will have different concerns than those working with learning disabled populations. Those who communicate on Tumblr will also have different styles and strategies than those who communicate on Twitter, and these will be different from those used in classrooms and museums. When first delving into new or social media, it is important to start by consuming content and becoming a part of some content community that is similar to what you wish to create.

This initial step of consuming content and joining into communities is one that often becomes a last step as well. It is very easy to jump into media and feel that no post should go unread and no Facebook post should go unacknowledged. Very quickly, people can lose their free time to the infinitely scrollable windows of constantly updating content, and platforms like YouTube are always ready with a new video to recommend. It is important to set both goals and limits going in. For instance, you could set a goal to follow diverse content creators who communicate about topics related to what you are interested in, and to grow your social media audience to 100 through interactions on relevant hashtags. You can also set a limit on how many minutes a day are spent looking at media. Users of the Pomodoro

(Cirillo 2006) method, for instance, might say they will only look at social media during one 5 min break per hour and new media during only one longer break a day. These kinds of goals and limits are transferable to almost all content areas, such as placing limits on screen time and video games, while creating goals to read more diverse authors. By starting with goals that are tied to engaging with existing content, you are setting yourself up to know what is good, what is bad, and to potentially join the community and build a reputation as a person before you try to establish yourself as an influencer.

As you grow comfortable with using the technologies, and once you are able to understand the day-to-day acronyms, language usage, and inside jokes of the platforms you want to use, it's time to create content. The language of the Internet isn't the same language used in formal writing and communications, and you will be accepted as a community member more readily if you use the language of the platform you are on, rather than formal language (Zappavigna 2012). As with consuming content, it's important to place limits on how long is spent creating content. Like all creative endeavors, perfect can become the enemy of good enough, and the creation of the perfect 3 min video or perfect instagram image can take hours if you let it. Your experience consuming content should teach you about the standards for your platform, and just as a rocket only needs to be nominal and not extraordinary, you only need to be good enough. The path to audience building is made by showing up over and over to create a steady stream of content, and it is better to have a myriad of videos than one perfect video, and a stream of tweets rather than that one perfect tweet that folks are likely to miss in their feed.

Audience building is a fickle process, and to a remarkable large degree, content quality won't bring in the eyes, ears, and minds by itself. To bring in followers you must promote your content through other means; tell people you are now on Twitter and promote your new Twitch channel in your traditional electronic newsletter. You also need your existing audience to share your content out to their audiences so that you can expand your reach by taking advantage of their reach. Finally, it helps to be lucky and/or famous. One clever tweet that happens to be retweeted by the right person or persons can go viral and may bring in thousands of new followers and millions of views. That exact same tweet, posted 5 min earlier or later, might not have been seen and shared the same way, and might not go viral. This is the aspect of luck, and it is difficult to try and create or replicate. Celebrity also plays a role that is often ignored. Companies like YouTube, Twitch, and Twitter will recommend prominent channels and accounts for new users to follow, and placement on these recommended lists will drive audience growth. Organizations with keywords in their names (e.g., NASA) will also grow due to the assumed authority that goes with their name, regardless of if they are or are not the best communicators on a given topic.

In determining success for your project, it is important to have reasonable metrics. For instance, if you will be developing content for your university or museum, strive to be as successful as your closest competitor who has the same media budget (e.g., the same amount of time and resources). You may want to be the number one voice in your speciality, but that may not be reasonable unless you are also a television host. Also, give yourself time—months not weeks—to achieve your audience engagement numbers. Bottomline: You can't grow instantly, and you can't compare yourself with someone with more fame or more resources. You can succeed however, through steady content creation as part of the greater community.

6.3 Best Practices That Are Here to Stay

It is in learning to use new and social media well that the real differences between these platforms emerge. Fundamentally, new media is a new way of pushing content to audiences, so special production techniques and technologies are required, but the core concepts from old media—narrative, pacing, and audio/video design—still apply. Social media, however, is its own unique form of communications, where not even the norms of spoken grammar necessarily apply!

6.3.1 New Media

There are two broad uses of new media in astronomy education: creating content that teaches astronomy, and having learners create content to demonstrate their learning of astronomy (and to ideally teach others).

Asking students to create content can be an excellent way to motivate students in non-majors classes to take a deep dive into a topic or topics. To the instructor, these videos are a break from the standard misery of grading essays that all follow the same structure, and for the students, this is a way for them to unite astronomy and other abilities in a fun and informative manner that teaches them astronomy can be found everywhere. For more advanced students, videos can be a way they can share their research in a digital elevator pitch and learn how to communicate in ways that make sense to the Facebook and Instagram generations. Need a starting point? Consider asking your students to create a science video that is a parody or tribute to something else, such as redoing music lyrics, or doing a science detective story in the style of Film Noir. While mandatory assignments shouldn't be required to go on the public Internet (see ethics section below), in-class creations can lead to out-of-class activities students volunteer or intern to do. The YouTube channel, "Active Galactic" is produced by students at the University of Arizona as an offshoot of their astronomy classroom experience. With 84 videos as of this writing, they have been produced by dozens of students who have used a remarkable diversity of styles to present content that ranges from interviews to music videos.

In creating content as a learning tool, there are countless possibilities. In deciding what to create, you need to answer three questions: (1) What need do I want to fill? (2) What content fills that need? (3) What can I do in a sustainable manner? Don't over think your answers. A need may be, "I want to force myself to keep up with astronomy news." That is a need. A need may also be, "I want to provide culturally-relevant astronomy lessons to native peoples through their phones." These needs are very different, but both can lead to highly-successful content creation that has an audience. In deciding what you can do in a sustainable manner, be honest with yourself. If you decide you want to create a series on the life of stars for little kids, you may have the ability to make high-quality animations in professional software, but not have the wherewithal to make more than two episodes. At the same time, it may be sustainable to do something just as educational with very simple animations created on an iPad.

In creating new media content, start with the basics: a story told with audio and/or video. Your production quality must be good enough that your audience doesn't have to put effort into understanding what they are hearing or seeing. Taking advantage of natural light in a quiet location is an excellent starting point for video (see Fraser Cain's *Guide to Space* series https://www.youtube.com/playlist?list=PLbJ42wpShvml6Eg22WjWAR-6QUufHFh2v), and the best quality audio can often be recorded while sitting in a closet filled with cloths (Impey et al. 2019). A modern camera phone can record passable audio or video, but an investment of a few hundred dollars can make a huge difference in quality. It is recommended that new content creators check out the latest "Best Equipment for …" articles to recommendations for their work. At the same time, the ability to record with a phone makes content creation something students can be asked to do for class.

It is hard to be more specific and say, "This is the winning recipe" because new media has proven over and over that there is an audience for anything. This is the long-tail phenomena (Anderson 2004). It was brought to light in this age of the Internet where companies often come to the forefront because they are able to sell a large amount of a niche product to a narrow audience. If people critique your idea, listen to the critique, but remember that all because something doesn't yet exist, that doesn't necessarily mean it shouldn't exist.

6.3.2 Social Media

Use of social media for informal astronomy education is straightforward and powerful. Use of it for formal education is a much more complicated minefield that is perhaps best summed up as: don't require students to use public platforms for classes, make sure you read your institution's social media policy before you mention your social media in a formal education setting, and check with your department chair or section head before you get started in case there are unintended consequences to your social media usage. At this time, we cannot recommend using public social media platforms in ways that require students to engage in those platforms (more on this in the ethics section below).

If you really want to use social media with your students, consider doing it in a simulated environment. For instance, have students plan out how they would schedule Tweets to live tweet a historic space event or discovery. Alternatively, have them imagine how a historical scientist would describe their life through Instagram. These kinds of lessons, coupled with information on how to follow real scientists and science discoveries through social media, can be both educational, and introduce students to ways they can be lifelong learners without having to look up from their phones.

When it comes to informal education on social media, the sky's the limit and the path is much less hazardous. There is a lot of low-hanging fruit and you can do good science education with nothing more than quick fingers and good use of search terms. From answering everyone asking "What's that bright thing next to the Moon?" to explaining meteor showers to everyone commenting on the shooting stars they're seeing, you can be the answer people need. This kind of search and respond creates teachable moments for as many people as you feel like messaging as well as their entire audience... and if your responses are shared, then the audience will just keep growing. With this kind of a simple intervention, you can directly engage people who have seen something or who have a known question. This "just in time teaching" isn't just educational, it also helps grow your audience and make you familiar to more and more people.

You can also bring your voice and your stories to prominent hashtags. Share your understanding of mission science during mission events or share your experiences when hashtags about doing science crop up. This too will allow you to speak to people seeking content: people who are searching on the hashtag because they are interested in its content. It also grows your audience and name recognition.

While much more difficult to grow, it's also possible to create niche accounts that serve a single purpose and don't respond to others or otherwise interact. An example of this kind of feed is @bitsofPluto (see Figure 6.1), which tweets small sections of Pluto's high-resolution mosaic. These kinds of feeds can grow by working behind the

Figure 6.1. The @BitsOfPluto twitter feed uses a bot by @hugovk to post a section New Horizon's Pluto imagery every 6 hr. Courtesy of NASA and Hugo van Kemenade @hugovk.

scenes to get prominent influencers to follow and share posts. When successful, these feeds are broadly educational and may be consumed by a wide audience.

Another strategy is to take a middle path and create a niche feed that posts in the voice of a project or mission, and responds to others and engages on hashtags as that entity. This is most commonly done by retail chains, and has become an artform for some fast food outlets. In science, this style of social media was first created by the GLAST (now Fermi) and Swift spacecraft on MySpace, and later perfected by Mars Phoenix on Twitter (see Figure 6.2). In this latter case, the mission even correctly lagged its responses so they would arrive as though the tweet being responded to had to travel from Earth to Mars and the response then traveled back again.

As with new media, the way you choose to use social media needs to be sustainable. Beyond that, as long as you follow the KISS principle—Keep It Simply Scientific—there is no wrong answer for doing astronomy EPO on social media. (You do not have to keep strictly to science, but the best practices of having a diverse feed are beyond the scope of this paper.) In general, accounts that are interactive are most successful, and accounts that mix information, dialogue on

Figure 6.2. The @MarsPhoenix account won numerous awards for its ability to emotional engage the Internet by Tweeting it's life and death from a first-robot perspective. Courtesy of NASA/JPL.

broader issues, and calls to action are the best at increasing engagement (Lovejoy & Saxton 2012, and references therein). These, however, are only guidelines, and new exceptions are created everyday.

6.4 Ethical Considerations

Working with new and social media opens content creators to liabilities they don't have in formal academic environments, or in most public education environments. While a museum may sell or share attendee data, it is unlikely that data will lead to anyone being cyber-stalked or harassed. The moment anyone puts data on a commercial server, they are at risk. As individuals, we can choose what risks we are comfortable with. As instructors, it is unethical for us to make choices for our students by mandating assignments that require content be placed on public platforms.

In addition to placing students into new risks, online accounts can also worsen existing risks. A student may be in a vulnerable situation that is managed by having no online presence. Beyond safety issues, there are also legal issues. Classroom instructors, unlike museums and other informal entities, must also abide by the Family Educational Rights and Privacy Act (FERPA, U.S. Department of Education 2019). In creating assignments, ask yourself, does this assignment, by its nature, break rules of FERPA by disclosing personally identifiable student information, such as an email address, to an outside entity.

While doing informal science education through these platforms is much less problematic, there are still ethical stumbling blocks that need to be avoided. Platforms such as YouTube and Twitch can be monetized and also allow donations. The ability to generate funding through these platforms is a boon in that it allows these activities to potentially pay for themselves. It also means that financial conflicts of interest can be created, especially if content is created using a personal account and promoted during work-related activities. Prior to monetizing content, it is important to consult with your appropriate institutional officials to make sure that you know how to not stumble into conflicts of interest.

Finally, it is important to remember that the fair use policies that apply in the classroom don't apply in the public arena. The Digital Millennium Copyright Act (U.S. Copyright Office 1998) protects the use of source materials in classroom settings for educational purposes as long as the use is confined to the classroom. This means that posting of materials in BlackBoard software, which requires a login to access, is protected, while posting of material on an instructor web page is not (U.S. Copyright Office 1999). By extension, posting broadly on the Internet is not protected. It is still possible to cite, link to, and otherwise reference content, and small excerpts for review purposes and that aren't seen to infringe on the ability of the copyright holder's ability to earn revenue, are still allowed. As always, use quotations and citations, and provide credits for all media.

6.5 Conclusions

While there is no one right way to create astronomy content for new and social media platforms, there are wrong ways, and there are better ways. The dos are simple: do set goals, do set boundaries, do what you can do sustainably, do (with exceptions) interact with others, and do seek out teachable moments using hashtags and search terms. The easiest ways to avoid problems are to follow the other KISS principle—Keep It Simply Scientific, and to avoid asking students to do classroom assignments on public platforms where their privacy can be at risk. Measuring success can be a challenge, and it is important to remember that fickle factors like fame and luck play a role and can't necessarily be planned. The best comparison you can make is to other entities in the same niche who have the same budget of time and resources. There is space in the new and social media ecosystems for all kinds of contributions, and by just showing up regularly with solid content, you can succeed.

References

Anderson, C. 2004, The Long Tail, Wired, https://www.wired.com/2004/10/tail/

Cirillo, F. 2006, The Pomodoro Technique (The Pomodoro), Agile Processes in Software Engineering and Extreme Programming, 54.2, 35

Impey, C, et al. 2019, Active Galactic Videos, https://www.youtube.com/channel/UCy0CQROnerc-bHhjFEjpHpw

Lovejoy, K., & Saxton, G. D. 2012, J Comput-Mediat Commun, 17, 337

U.S. Copyright Office 1998, Executive Summary, Digital Millennium Copyright Act, https://www.copyright.gov/reports/studies/dmca/dmca_executive.html

U.S. Copyright Office 1999, Copyright and Digital Distance Learning, https://www.copyright.gov/docs/regstat52599.html

U.S. Department of Education 2019, Family Educational Rights and Privacy Act (FERPA), https://www2.ed.gov/policy/gen/guid/fpco/ferpa/index.html

Zappavigna, M. 2012, Discourse of Twitter and Social Media: How We Use Language to Create Affiliation on the Web, Vol. 6 (New York: Bloomsbury)

Chapter 7

Education Through Exploration: A Model for Using Adaptive Learning to Teach Laboratory Science Online

Chris Mead, Ariel D Anbar, Lev B Horodyskyj and Donald Bratton III

Adaptive learning technologies are becoming more common and they have the potential to transform the way laboratory science is taught online. In this chapter, we introduce adaptive learning, discuss the research supporting its effectiveness, and list prominent technology providers of adaptive learning. To ground this concept, we introduce the Education Through Exploration (ETX) model of digital learning design, which leverages adaptive learning and is specific to the sciences and laboratory science in particular. We use examples from our own work to illustrate both adaptive learning and the ETX model.

Chapter Outcomes

By the end of the chapter, readers will:
- Understand what adaptive learning is and why it is effective.
- Understand the Education Through Exploration model and its research foundation.
- Be able to start conceptualizing their own adaptive learning designs.
- Have a list of educational technologies that provide adaptive capabilities.

7.1 Introduction: What Problem Are We Solving?

Among the many reforms to science education that have been made in recent years is the push to make science laboratory activities resemble authentic scientific practices (e.g., Hofstein & Lunetta 2004). Laboratory teaching sections and activities have long existed to teach practical skills associated with a scientific discipline and to demonstrate the implications of concepts taught elsewhere in a class, whether through lecture, readings, or otherwise. Recently, there has been a move to shift the emphasis of such laboratory activities to demonstrate and reinforce "scientific habits of mind" and "understanding of the nature of science" (Hofstein & Lunetta 2004).

In practice, classroom laboratories fall between two end-members: the verification or "cookbook" lab and the authentic, inquiry-based lab. Inquiry learning has a long history, including the Learning Cycle (Karplus & Thier 1967; Lawson 2010) and its extension, the popular 5E model (Bybee et al. 2006). Even though inquiry-based teaching is demonstrably more effective across various metrics, it is not yet universally used (Brownell et al. 2012; Pearson et al. 2010; Blanchard et al. 2010; Hofstein & Lunetta 2004). This lag between best-practices recommendations and on-the-ground use stems from teachers' lack of awareness of these practices and their lack of expertise and comfort in using them.

As much as this gap is a challenge for in-person teaching, it is still more challenging for online teaching. As college and even high school courses have moved online, they have too often taken on the worst features of existing in-person teaching practices. In particular, this means instruction that relies on video lectures and simple, computer-graded quizzes (Koedinger et al. 2015; Toven-Lindsey et al. 2015). Although there is an overreliance on these kinds of passive learning in traditional classroom settings, online delivery further encourages their use because students are commonly learning asynchronously from their instructor and peers. To provide active learning online requires a system in which students' choices and inputs receive meaningful feedback. In this chapter we will discuss the concept of *adaptive learning* and the ways in which it can be used to provide this kind of feedback and thus enable effective active learning online.

Adaptive learning refers to a range of technologies that deliver a dynamically personalized learning experience to each student based on the student's right/wrong answers or their stated or inferred interests (Shute & Zapata-Rivera 2007; U.S. Department of Education 2013). By using these adaptive learning technologies, it is possible to design and build inquiry-based labs that can be delivered online for asynchronous use. The authors of this chapter and their collaborators have used this approach to create two online lab science courses, both drawing on astrobiology concepts, as well as a number of standalone inquiry-based lessons in astronomy, earth science, and other fields.

This chapter will have three main sections:
- An introduction to the concept of adaptive learning.
- A proposed instructional design model for the use of adaptive learning to create inquiry-based science labs.
- Examples of this model's use in astronomy and astrobiology education.

7.2 The *What* and *Why* of Adaptive Learning

Adaptive learning and related terms such as personalized learning and intelligent tutoring refer to various ways in which student data can be used to inform the instruction that is offered and to improve learning outcomes in computerized environments (Shute & Towle 2003; Shute & Zapata-Rivera 2007; VanLehn 2011; U.S. Department of Education 2013; Mavroudi et al. 2018). Adaptive learning as defined in this way has been shown to be more effective than comparable non-adaptive designs (VanLehn 2011; Ma et al. 2014; Kulik & Fletcher 2016). This section will touch on why adaptive learning is more effective and introduce some of the most readily available methods for adaptive learning.

The benefits of adaptive learning are best examined through the lens of a constructivist theory of learning (Fosnot & Perry 2005). Constructivism holds that learning is not a simple linear process, nor is the process of learning a given topic the same for different people. This is because each student must construct his or her understanding on a framework of their prior knowledge and experiences. That is not to say that the student must do this alone, however. A tutor, a peer, or in this case, an intelligent tutoring system can play a key role in helping the student build an accurate and complete understanding of a topic.

Learning with a tutor is more effective than learning independently for several reasons. The tutor can encourage metacognition, prompting the student to review material or test their understanding. The tutor can help diagnose errors and recommend new strategies or approaches. The tutor can also note connections across topics, opportunities for further learning, or next steps. The benefits of human tutoring have been well-studied (Bloom 1984; Cohen et al. 1982; Chi et al. 2001). More recently computer tutoring has been shown to have similar effectiveness under certain conditions (VanLehn 2011; Ma et al. 2014; Kulik & Fletcher 2016).

Although many different types of adaptivity exist, our discussion and use of it will be mostly limited to adaptivity that responds to a student's current actions, including what a given answer implies about the student's current content knowledge or what a given problem-solving strategy implies about the student's procedural knowledge (Figure 7.1(A)). This is in contrast to the more complex, algorithmic adaptive learning systems, which build a detailed *learner model* and use machine learning or expert systems to select an appropriate learning activity for each learner at each

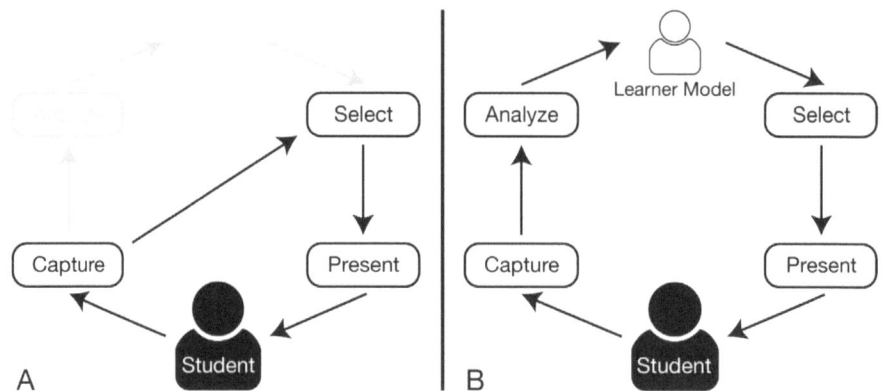

Figure 7.1. Schematic diagram of how adaptive learning functions (Shute & Zapata-Rivera 2007). The *student* performs actions within a lesson. Each of those actions are *captured*. In simpler adaptive learning designs (Panel A), those actions are used to *select* the next learning activity or content which is *presented* to the student. In more complex systems (Panel B), the student actions are *analyzed* using a *learner model* to determine what activity or content the student is presented with.

point in time (Figure 7.1(B)). Although these algorithmic systems are quite powerful, their use to date extends to only a subset of science topics. Moreover, creating a new course with such a system requires extensive resources to build and train the underlying computer models.

The remainder of this section will explain the types of student needs adaptive learning can address, how students who require support are identified, how this support is provided, and what adaptive learning looks like in an inquiry-based science lab. See the final section of the chapter for links to specific adaptive learning technology providers.

7.2.1 What Student Needs Can Be Addressed?

The most straightforward needs are those where a student only needs a reminder about a concept or procedure that he or she has learned before. This could be a formula, a definition, a classification scheme, or the order of a set of steps. We call these straightforward because we are assuming that the student already understands the concept and requires minimal instruction.

More challenging is the situation where the student has not learned some piece of expected prior knowledge. In this case, as opposed to the first, simply providing the formula may not address the knowledge gap—the student here needs *instruction* on the topic, not just a reminder. Similarly, a lesson or curriculum could be structured so that students encounter new topics in an unpredictable order. In this case, adaptivity could provide the appropriate instruction as needed based on the path each student takes through the lesson. This category also includes the issue of misconceptions, where the student thinks they understand the concept, but actually has inaccurate or incomplete knowledge.

Alongside these content-based needs, students may also need guidance or feedback about their approaches to studying, learning, or assessment. This could include not spending enough time or repeating the same approach to a problem through multiple failures instead of thinking of an alternative. Student needs related to metacognition, self-regulation, and other issues can be significant barriers to success, so adaptive designs have great potential here.

7.2.2 How Are These Needs Determined?

In the most basic type of adaptive learning, students might have the option of choosing to receive extra help or information, e.g., "Click here for a refresher on logarithms." This is essentially non-automated adaptive learning and it can be desirable for optional or supplemental material. Using an adaptive design in this case allows students to choose how much they want to learn about the subject above and beyond the minimum amount that is required by the instructor. This design is not as well suited for gaps in knowledge, because the student (a) may not realize that they have a gap in knowledge (low metacognition) or (b) may not choose to spend the time to address this gap in knowledge (low self-regulation). A compromise, however, is to offer the supplemental material at the start of the lesson while also using automated adaptivity to detect students who struggle later in the lesson and who may benefit from the material they skipped at the start.

Automated adaptive learning infers gaps in skills or knowledge from the student's answers and other interactions with the learning environment. This can be done in a number of ways and with varying complexity:

- Adaptive content can be triggered in response to one or more specific responses to a single question. For example, if the answer was off by a factor of 1000, the feedback might suggest that the student forgot to convert meters to kilometers.
- On a more complex question, detailed and specific adaptive content can be triggered in response to a combination of particular wrong (or right) answers. For example, if a student had two answers that contradicted each other, the feedback could call attention to that without necessarily indicating which (if any) of the current answers was correct. These small corrections are helpful, but still keep the onus on the student to work through the problems on their own.
- There can also be another level of different adaptivity depending on the pattern of responses to the same question. For example, a student whose responses suggest a trend toward improvement can receive different feedback than a student who is repeating the same mistakes.
- The most complicated kinds of adaptive learning build a multi-dimensional model for each student and use that to predict what kind of activity will be most useful for that student at that current point in time. For example, in introductory physics, the adaptive learning system might determine that a student has mastered problems involving positive acceleration, but still struggles with negative acceleration. In response, the student could receive targeted practice on that topic.

7.2.3 How Is Adaptive Learning Delivered?

When it comes to the mechanism for how adaptive learning is delivered, we will introduce two broad categories: *adaptive feedback* and *adaptive pathways*. Adaptive feedback is delivered on the same "screen" as the question or activity that triggered the feedback and it typically aims to address one or more mistakes related to that question. This kind of adaptive design is appropriate when it is assumed that the student has made a minor error that can be easily corrected, such as the reminders described above. Adaptive feedback can also be useful when students work with detailed interactive simulations, because the feedback can responsively guide students to a correct solution, even through multiple small mistakes.

If it is likely that the student has a broader misunderstanding, gap in knowledge, or a durable misconception, then an adaptive pathway is the better design. Adaptive pathways are sets of questions and instruction that will only be shown to some students—those who meet some predetermined set of criteria and who are expected to benefit from this additional instruction. Although adaptive pathways can be brief and specific to the current lesson, this category also includes larger digressions and can even route a student back to material from previous lessons.

7.2.4 What Does Adaptive Learning Look Like for Inquiry-based Science Labs?

As noted in the introduction, there is a strong desire for greater use of authentic science in laboratory science courses. Inquiry-based laboratory courses are intended to help students learn scientific practices and ways of thinking. To be true inquiry, a lesson should offer the student some autonomy in identifying the question to investigate, selecting the methods or approaches to the investigation, and interpreting the results of the investigation. The obvious challenge in asynchronous online settings being that students who exercise such autonomy may struggle or get stuck and be without rapid feedback to get back on track. An adaptive inquiry lesson allows students to have autonomy to make a range of decisions. Students can take what turn out to be unproductive approaches to solving a problem, and adaptive feedback and pathways can be designed to help them *at the appropriate time*. That is, the adaptive design can allow students time to recognize and learn from these failed attempts.

7.3 The Education Through Exploration model

We began this chapter by talking about the need for online, inquiry-based science labs, the challenges inherent in providing them, and the potential for the techniques of adaptive learning to meet those challenges. In our own work we have designed and built a number of online, inquiry-based labs, from which our research and development team has defined a development model called Education Through Exploration (ETX). The core of this model is that curiosity—the desire to understand new things—and the satisfaction of discovery should be used to motivate students to master the skills of exploration, i.e., scientific inquiry (Figure 7.2). This approach to design and these courses fundamentally depend on adaptive learning. Without adaptive technology, it would be very difficult to provide students with enough support to work through meaningful problems in a fully-online environment.

The pedagogical ideas underlying the ETX design model are well-established in the conventional science education literature. The Learning Cycle (Karplus & Thier

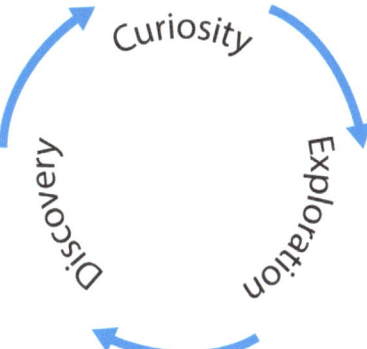

Figure 7.2. ETX learning loop. Curiosity drives exploration. Exploration leads to discovery. Discovery inspires further curiosity.

1967; Lawson 2010) consists of three phases: exploration, term introduction, and concept application. The popular 5E model (Bybee et al. 2006) extended this original Learning Cycle to include engagement and evaluation components. The Learning-for-Use model (Edelson 2001) describes somewhat similar steps: motivation, knowledge construction, and knowledge refinement. The ETX design model makes two substantive contributions to this body of thought. First, the ETX design model proposes the idea that curiosity is important not just as motivation, but an important outcome itself. Second, the ETX design model incorporates digital learning technologies, which changes what is possible in the space of inquiry-based learning and makes it far easier to inspire curiosity in students.

It is important to clarify that inquiry in ETX designs is guided inquiry, not open inquiry. Here, open inquiry refers to an activity in which students, not the instructor, generate the question and method of investigation in addition to being responsible for the ultimate interpretation of the results (Blanchard et al. 2010). Although open inquiry can be effective, researchers have criticized it as being less effective than guided inquiry or even non-inquiry (Klahr & Nigam 2004; Kirschner et al. 2006; Settlage 2007; Blanchard et al. 2010). In addition, for our purposes in building asynchronous online learning experiences, the difficulty of accounting for complete student autonomy in an open inquiry design makes it infeasible.

To design learning experiences that create the learning loop shown in Figure 7.2, we have identified a number of *design practices*, organized into three categories: *conveying authentic science*, *learning as a journey*, and *digital by design*. Our goal is that these practices lead to learning experiences in which an interesting setting, phenomenon, or question inspires curiosity to motivate students to explore; to experiences in which that exploration naturally takes the form of asking and answering scientific questions using observations and data; and to experiences in which answering those scientific questions inspires new and interesting questions for future explorations. These categories and practices are best thought of as heuristics, so there is some overlap among the categories and the exact use of the practices will vary from lesson to lesson, as illustrated in Section 7.4 of this chapter.

7.3.1 Conveying Authentic Science

The practices within *conveying authentic science* encompass the essential qualities of inquiry learning as well as features designed to spur curiosity and interest.

Most importantly, students using an ETX lesson should learn science by doing science, that is, by engaging in scientific investigation. This means that the students' activities should be motivated by a question or hypothesis, that students should be able to collect or compile data, and that students should apply scientific reasoning to draw a conclusion about the guiding question based on the available data. To be consistent with true inquiry (see Blanchard et al. 2010), ETX lessons should include opportunities for student choice, such as selecting a method of observation, order of investigation, data quality threshold, etc. Students should also be able to interpret the results of the investigation.

Another important aspect of an ETX design is the use of real-world and interdisciplinary problems. These related ideas provide both educational and motivational benefits. The complexity of real-world problems demands that students understand concepts in multiple contexts. Similarly, integrating concepts from multiple disciplines provides a better illustration of how science is used to solve complex problems than the traditional siloed introductory science course. Working on real-world problems can also be more compelling than working on abstract examples, particularly for students who are not majoring in science. In some cases, this could even be an opportunity for students to write or think about connections to their own major and how the scientific issue under discussion might be relevant beyond the immediate inquiry task.

Last, as much as possible these learning experiences should connect to the frontiers of knowledge in a given field. This is not only a demonstration of authentic science, but by posing questions without clear or certain answers there is an opportunity to discuss how science deals with uncertainty and how an unknown, but constrained, answer can still be useful. Moreover, in a research university setting, professors can look for examples from their own research.

The goal of conveying authentic science is supported by techniques from our third category—*digital by design*. It also provides useful constraints to our second category—*learning as a journey*—in that the process of reaching a final answer should be of equal importance to reaching the correct answer and that the assessments and adaptive feedback should reflect this ideal.

7.3.2 Learning as a Journey

The practices within *learning as a journey* serve to create learning experiences whose structure and flow support the ETX learning loop (Figure 7.2). As the name implies, an ETX learning experience should emphasize growth in knowledge and capabilities. It should demand an appropriate level of mastery, but the current goal(s) should be clear to the student and failure should always be used as an opportunity for learning.

Communicating the learning objective of an experience to the students allows them to monitor their progress and to better connect what they are learning to what they already know. This is important in an ETX learning experience too, but in addition to high-level learning objectives, we emphasize the goals of the scientific problem-solving activity that are specific to the lesson and those of the course as a whole. Including these tangible goals in addition to the more abstract learning objectives grounds those objectives and gives the students stronger motivation toward achieving both types of outcomes. The nested short and long term goals also serve to link concepts together across the course, which helps to cement understanding (cf. distributed practice). As an example, the "Marble Bar" virtual field trip in *Habitable Worlds* takes students to a visually complex rock formation in Australia. The lesson begins by asking students to discover "how this rock sequence formed." To move forward in the lesson, students zoom in to make a series of more small-scale discoveries before zooming out at the end to answer the larger, initial question.

ETX designs have numerous opportunities for students to learn from failure. Failure at a task in a learning environment is often interpreted negatively (by students, but also perhaps by teachers). *In actuality, failure is almost always a point at which learning can occur* (e.g., Kapur 2008). In ETX designs, or other kinds of adaptive learning, the use of adaptive feedback and adaptive pathways is a powerful tool for helping students to learn from failure. Most clearly, these instances are opportunities to better understand the specific concept or procedure around which the mistake was made. There are also opportunities to learn metacognitive skills from failure. Here, adaptive designs are even more powerful, because the system can track behaviors and responses to failure through time for each student. For example, a student who makes the same mistake several times in a row could benefit from reflecting on his or her strategies rather than just checking surface-level problems.

Related to learning from failure and consistent with the ideas in conveying authentic science, ETX designs emphasize the process and the steps required to solve a problem just as much as the ultimate answer to that problem. That is not to say that the answer is not important, but rather that reaching an answer without following a good process misses the point. In practice, this means that ETX designs might pose a conceptual question instead of or in addition to a calculation question. It also means that the choices and considerations of an experimental design should be a part of the learning experience.

7.3.3 Digital by Design

Cutting across all of these practices is the use of digital technologies, including visualizations or other multimedia, interactive simulations, and intelligent tutoring. The phrase *digital by design* is meant to imply that these designs are not constrained by the limitations of in-person learning. The starting point for a new digital learning experience design should be "what is the best way for a student to understand this topic?" In contrast, an in-person design necessarily begins with a compromise, namely, "what is the best way for a student to understand this topic *that can be done in a classroom?*"

The ideal activity to learn a specific concept may not be possible in an in-person class. Perhaps it is logistically challenging, expensive, or impossible. Perhaps it would take days, years, or millennia to experience in real-time. Any one of these can be overcome in a digital learning environment. This recommendation also means that an existing in-person learning experience need not, or perhaps should not, be the starting point when designing a digital learning experience. If the digital version can do something novel that is better, it should.

Active learning—the idea that students learn best when they are cognitively engaged—is not at all unique to digital learning, but it remains just as important in that realm as it is in-person. In the context of *digital by design*, active learning is both a trap and an opportunity. To call back to the chapter's introduction: it is easy to create passive digital learning experiences. However, digital learning also offers compelling and unique ways for students to learn actively, and this is the opportunity. Following from the other practices, and particularly *conveying authentic science*, ETX designs employ active learning through their focus on scientific exploration and discovery. Having access to a range of assessment types (simple questions, simulation-based questions, etc.) also creates more opportunities for students to learn actively.

Finally within *digital by design* is the use of adaptivity. Of course, this topic does not need special treatment in this particular chapter, but we will conclude this section by noting that while some aspects of digital learning go beyond what is possible in-person, there are some aspects of in-person learning, such as being able to receive immediate expert feedback from an instructor, that digital learning has only begun to replicate.

7.4 Examples of ETX Model and Adaptive Learning in Online Astronomy and Astrobiology Education

Using ETX designs, we have developed and deployed two adaptive, fully online undergraduate-level courses emphasizing astrobiology. In this section we will use these to illustrate the techniques described in the previous section. These examples were all built using the Smart Sparrow Adaptive eLearning Platform (Ben-Naim 2011). Smart Sparrow provides adaptive learning capabilities that can be built and modified entirely within a browser-based lesson authoring environment. Rapid end-user editing and customization has been important to the development of all of the examples we will show. All of the examples we will discuss are web-browser-based.

7.4.1 Habitable Worlds

Habitable Worlds is an online astrobiology course intended primarily for undergraduate non-science majors. At Arizona State University, it fulfills the quantitative science general education requirement. Unlike most traditional courses meeting this requirement, which have separate laboratory and lecture sections, *Habitable Worlds* has a unified learning experience that blends applied thinking and active problem solving throughout all lessons. Although it does include some short "lecturette" videos, these are primarily used to bookend the active learning. Additional information about *Habitable Worlds* can be found in Horodyskyj et al. (2018).

The course scaffolds the student experience narratively and intellectually with the goal of finding a habitable planet beyond our solar system. This goal is systematized and explained through the famous Drake equation (Drake 1961). The terms in the Drake Equation are used to introduce and motivate new topics. For example, the $R*$ term—the average rate of star formation—motivates learning about the properties of stars and the process of star formation, while the n_e term—the fraction of planets that are "Earth-like"—reveals the need to understand what an Earth-like planet is and in what circumstances one might exist. The course rewards mastery of the concepts related to each Drake equation term in the final project, where students must use their knowledge to find at least one of a handful of habitable planets orbiting 500 fictional candidate stars.

Habitable Worlds delivers inquiry learning experiences primarily through the use of interactive simulations, immersive and interactive virtual field trips (iVFTs), and basic inputs like checkboxes or multiple-choice questions. Adaptive feedback is used throughout the course. Any page with student input is programmed with multiple triggers to respond intelligently to students' activities on a page, which can range from simple numerical inputs to complex simulator set-ups. The interactive components are built so that actions within a simulation or iVFT can be "seen" and responded to by the adaptive learning system. This is critical to our ability to build meaningful inquiry learning in *Habitable Worlds*.

We will discuss two examples from *Habitable Worlds*. The first, a lesson about the properties of stars, reflects design elements used in numerous lessons across the course. The second, the course's summative, multi-week, final project, is an example of the novel approaches that are possible with fully computer-based instruction.

7.4.1.1 The Properties of Stars
The brightness–distance activity early in the course illustrates a number of the techniques described in Section 7.3 as well as how adaptive learning is used in *Habitable Worlds*. This lesson introduces the brightness–distance relationship, which is key to understanding later concepts including stellar luminosity and other properties derived from the luminosity. Figure 7.3 shows screenshots from an investigation in this lesson.

Although this lesson does not make use of cutting-edge science, it does provide students a chance to develop a foundation for scientific investigation. The lesson trains students to test a hypothesis for basic plausibility before moving forward. Here, if students propose that more distant objects are *brighter* than or the same brightness as closer objects, they are sent down an adaptive pathway that asks if this hypothesis is consistent with what they know from direct experience (Figure 7.3(B)). Later in the investigation, students are given the freedom to propose a methodology for collecting data. The adaptive design ensures that their methodology meets certain criteria, but otherwise students have agency over setting up their methodology. With adaptive feedback, we can even hold students to their methodology, so if they said that 20 data points were needed they must collect at least 20 before progressing. The predict–observe–explain cycle shown here is used across *Habitable Worlds* in various forms.

Another design highlight from this example is the way that a digital learning experience can provide instant feedback and allow the student to try several alternative solutions quickly. Apart from the fact that collecting apparent brightness data in-person would be difficult, overlaying multiple functional forms of the brightness–distance relationship would be slow and could distract from the actual learning objective. In addition, the next phase of the lesson allows the student to test whether the brightness–distance relationship for our solar system applies to stars of different luminosity—to test for the generalizability of their explanation. This is another important scientific principle and one that the digital, adaptive design readily supports.

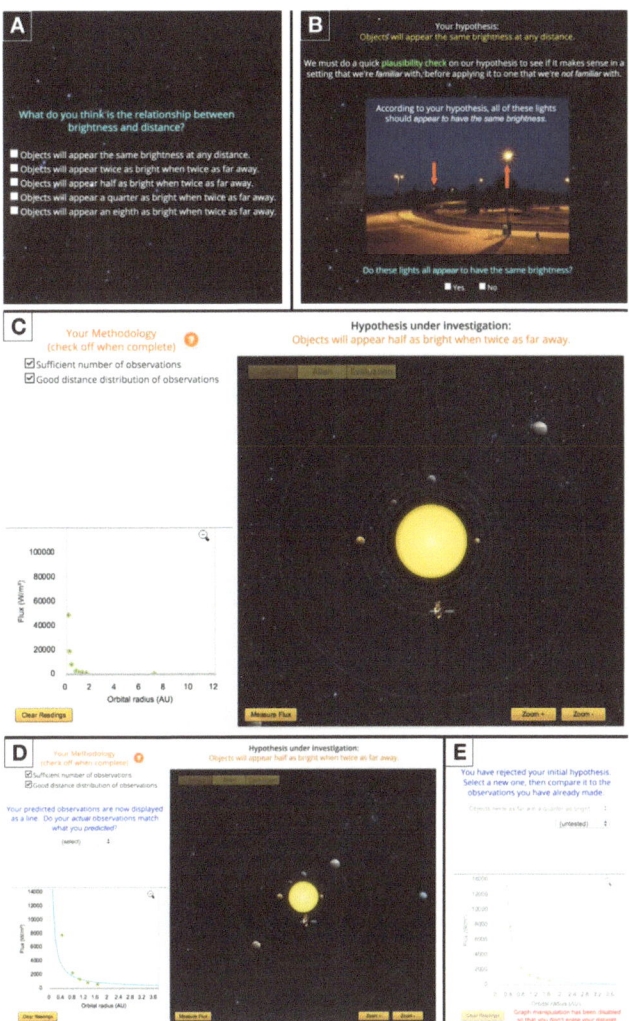

Figure 7.3. Screenshots showing a predict–observe–explain cycle in the second $R*$ lesson. Students make a prediction about the mathematical relationship between distance from a light source and its apparent brightness (A). Adaptive feedback is used to check that this prediction is plausible (B). They then use a solar system simulation to plot this same relationship (C) and compare these observations to their initial prediction (D, prediction is overlain as a line). The cycle concludes with the student evaluating their prediction and identifying the true relationship if their initial hypothesis was incorrect (E).

Finally, through this example we can illustrate how adaptive learning designs can be iteratively improved. The brightness–distance activity has undergone a series of improvements that were informed by study of student learning analytics data collected by the Smart Sparrow platform. We reviewed the amount of time spent, number of attempts made, and patterns of wrong answers chosen by students. The first redesign sought to reduce the time and number of attempts spent on the "experiment" section by adding more detailed adaptive feedback and pathways to

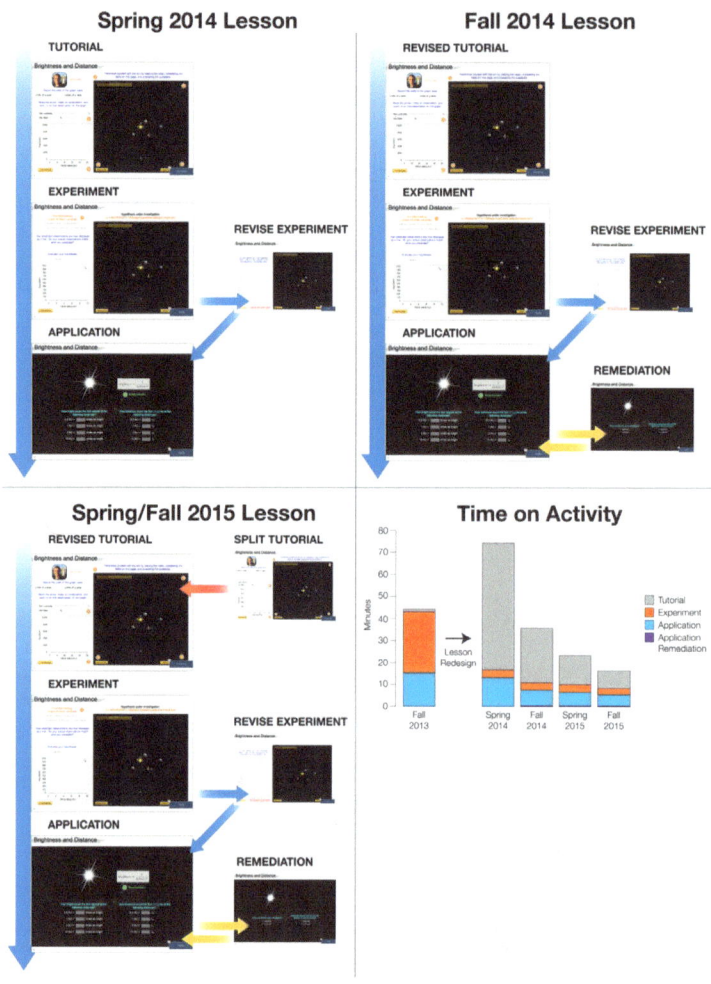

Figure 7.4. Panels illustrate the brightness–distance lesson structure across multiple revisions, beginning with the initial redesign in Spring 2014. The last panel shows how the length of time required to complete the lesson and its subsections changed following each revision. The final version successfully reduces the time spent on the experiment section without increasing the overall time required to complete the lesson.

the preceding tutorial section. As the graph in Figure 7.4 shows, the Spring 2014 change did reduce the time students spent on the experiment, but at the cost of much more time spent on the tutorial. Subsequent improvements in Fall 2014 and Spring 2015 reduced the total time required to complete the lesson to below the Fall 2013 baseline. Note that the learning objectives and summative activities were essentially unchanged across these revisions.

7.4.1.2 Habitable Worlds Final Project
In the *Habitable Worlds* final project, students are tasked with searching 500 candidate stars to find a habitable planet. To accomplish this, they must apply the

skills they have learned from the course to complete calculations, build models of their star–planet system, and identify habitable worlds before the end of the term. Some of the project interface views are shown in Figure 7.5. Students are graded on their work in two ways: data quality and "scavenger hunt" objects found. These grades reflect the accuracy and appropriateness of calculations and the breadth of study, respectively. This summative activity not only tests for understanding of concepts from throughout the course, it also serves as a narrative endpoint for the course and represents a tangible accomplishment for students that demonstrates their achievement.

Because the project serves as the course's final assessment, we do not provide adaptive feedback or pathways. However, a modified version of the course might

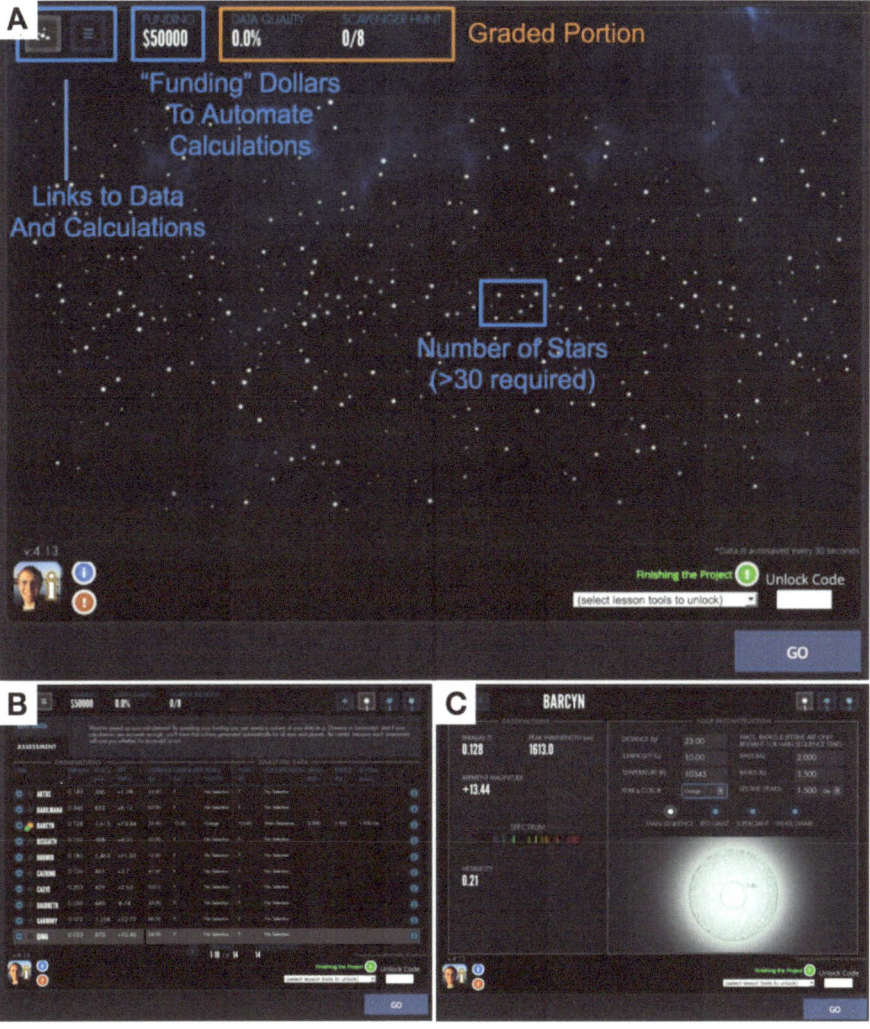

Figure 7.5. Screenshots from *Habitable Worlds* final project.

choose to more tightly scaffold this activity by building adaptive pathways that led back to the earlier lessons in which students first learned the concepts required by the project. The current version does offer some adaptive support by allowing students to use tools (simulations or equations) from elsewhere in the course. To unlock these tools for use in the project, students must have completed those earlier lessons, which serves to gate each student's progress to material that they have already learned.

7.4.2 BioBeyond

Like *Habitable Worlds*, *BioBeyond* is an online biology course, borrowing on astrobiological themes, intended for undergraduate non-science majors. Rather than learning how to locate habitable exoplanets, *BioBeyond* students learn about the processes of life and evolution, how life shapes and is shaped by geology, and what it means for a planet to be habitable. *BioBeyond* was designed and built via collaboration between Smart Sparrow and Arizona State University.

There are additional differences between the courses, some of which may be instructive for other adaptive learning designers. *BioBeyond* was designed to meet the common requirements for introductory biology, yet within these external requirements there was still room for creativity such as the decision to focus on astrobiology. *BioBeyond* often uses historical scientific discoveries as a narrative and scientific framework for its inquiry activities. This grounds the activities and could motivate supplemental activities related to those scientists or discoveries. Finally, although both courses employ active learning extensively, the design of *BioBeyond* uses interactivity on nearly every screen, adding in more opportunities for students to demonstrate understanding or check their work.

Our example from *BioBeyond*, the Time Traveler's Guide, showcases immersive, interactive virtual field trips (iVFTs). This technology is also used to a smaller degree in *Habitable Worlds*.

7.4.2.1 Time Traveler's Guide
The Time Traveler's Guide unit in *BioBeyond* makes extensive use of iVFTs and offers a good illustration of both adaptive learning and the ETX design practices. The premise of this unit is that the student can travel back in Earth's history in order to observe what organisms existed and what types of environments were present at three key points in time: 65 million years ago, 560 million years ago, and 3.5 billion years ago (Figures 7.6 and 7.7). The students make initial predictions about what they will find at each temporal destination and then use iVFTs to explore the three different environments. Using the adaptive learning platform, it is simple to remind students of their initial prediction and ask them to revise it once they better understand each paleoenvironment.

iVFTs are an excellent tool for active, inquiry learning. We discuss this digital learning technology and its application at greater length in Mead et al. (2019). Most obviously, iVFTs bring students to places that are difficult or impossible to access in-person and, for online learning, they bring the benefits of in-person

Figure 7.6. Screenshots from the Time Traveler's Guide iVFT set 65 million years ago. The first panel shows how students are helped through basic navigation in the virtual environment. The second panel shows one of the ways students can collect data about the field site. The third panel shows a formative assessment. Note the adaptive feedback in the lower right.

education to the digital realm. In Time Traveler's Guide, students explore field sites in Western USA, and South and Western Australia, crossing not only geologic history but also the globe. Within these diverse and interesting sites, students are able to learn by applying simplified, but realistic scientific methods, including data collection. In the 560 million year ago lesson, for example, students find fossils in a rock outcrop and then use a dichotomous key to uniquely identify each organism represented.

Figure 7.6. (Continued.)

7.5 Implementing Adaptive Learning

Adaptive learning designs, though not new, are far from the norm in online or computer-based learning. The slow adoption of adaptive learning is in part simple inertia—any change is difficult. It is also due to the real and perceived challenges of implementing adaptive designs. Our goals in writing this chapter were to provide instructors and other educational decision makers with the background information necessary to decide if adaptive designs meet one of their current needs and to provide these potential users with a foundation from which they can create or customize new adaptive learning designs. In this final section, we will briefly discuss some practical issues associated with implementing adaptive learning.

There are many educational technology companies that offer adaptive learning products. These companies and products each take a different approach in providing an adaptive learning experience to students. They also vary in terms of the subject diversity of available pre-made courses as well as how much control instructors have over making changes to those courses. Some well-known products (listed alphabetically) include:

- Acrobatiq: http://acrobatiq.com/
- ALEKS: https://www.aleks.com/
- Cerego: https://www.cerego.com/
- CogBooks: https://www.cogbooks.com/
- Knewton: https://www.knewton.com/
- Open Learning Initiative: https://oli.cmu.edu/
- Smart Sparrow: https://www.smartsparrow.com/

Figure 7.7. Screenshots from the Time Traveler's Guide iVFT set 3.5 billion years ago. In the first two panels, students are visually introduced to the ancient fossils that can be found at this field site. The third panel shows the use of Gigapan imagery, which allows students to explore a large fossil bed up close and search for the kinds of fossils they learned about previously. The fourth panel illustrates a drag and drop response format in which students compare the fossilized stromatolite texture to various modern rock textures. Note the adaptive feedback in the lower right.

The website EdSurge also hosts a tool that allows instructors to filter more than 50 current courseware products by features, including adaptive learning. This is found at: https://www.edsurge.com/product-reviews/higher-ed/courseware.

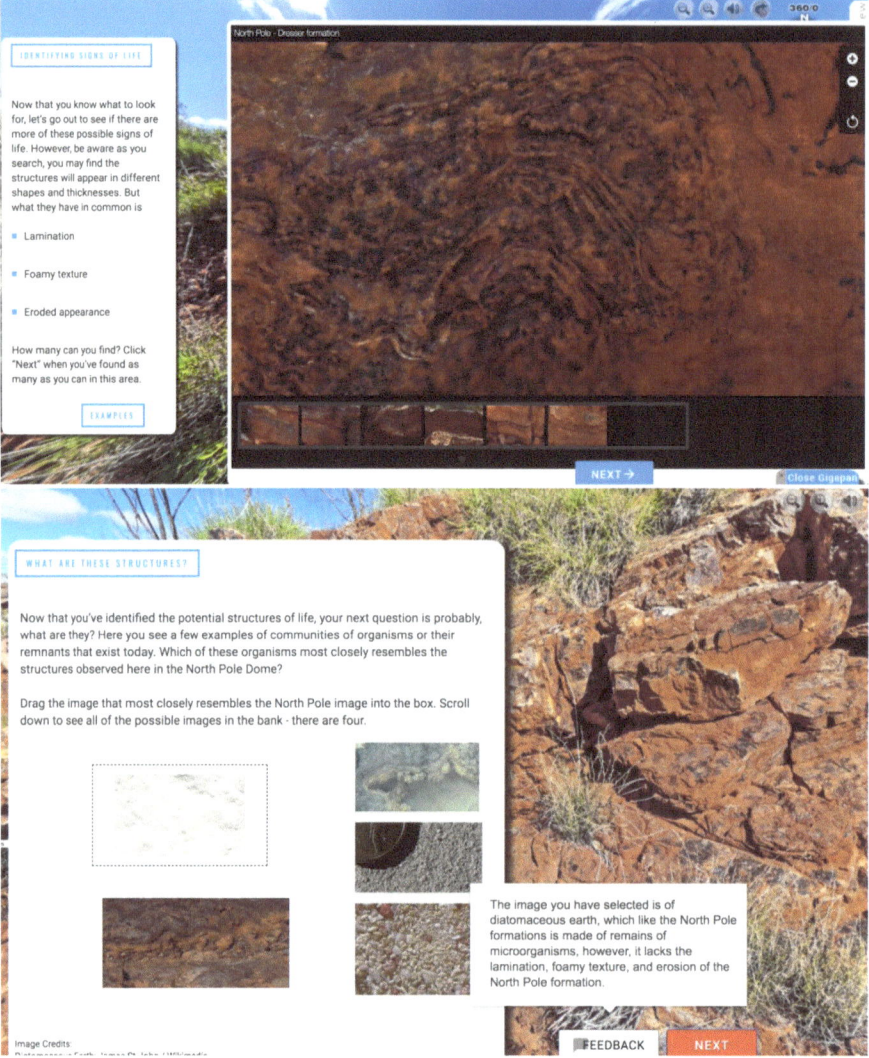

Figure 7.7. (Continued.)

In addition to varying by subject matter and focus, these existing products also vary in other ways. In some products, most or all instructional material is selected algorithmically—that is, that every student may receive a different set of tasks to complete. Others rely more on the learning designer to build a structured lesson, with adaptivity being used to support students along that path. Some products include a course "author," allowing local users, such as the course instructor, to make changes or even create new materials. Last, the use-cases for these products will differ, so it is important to consider whether a particular adaptive learning product can be used as a course supplement for an in-person course—as homework or as classwork in a flipped mode—and/or if it can be used as a standalone course delivered fully online.

Numerous other practical considerations exist. Although we cannot address them in detail, we will note them here. First, it is typically possible to connect an adaptive learning product to learning management systems (LMS), such as Blackboard or Canvas. Doing this not only simplifies the process of assigning lessons and recording grades, the use of "single sign-on" makes it easier for students to verify their credentials and gain access to the lessons. Second, although web-browser-based products are very convenient, when technical issues do arise, they can be very frustrating for students. It is necessary to ensure that students have reliable options for technical support, whether through the school or through the adaptive technology company. Finally, while adaptive learning is powerful and effective, there remains a need for an active instructor presence, particularly in fully-online courses. From our experience with *Habitable Worlds*, the discussion board helped instructors to identify common content issues as well as to form a personal connection with students in the class.

7.6 Conclusion

This chapter has introduced the concept and explained the fundamental mechanisms of adaptive learning technology. It also presented the Education Through Exploration design model, which relies heavily on adaptive learning, as one approach for offering meaningful and effective inquiry science learning online. The use of adaptive technology for science learning is in its early days and, therefore, far from reaching its full potential. Importantly, the fundamental tools and approaches for designing adaptive learning are certain to evolve and improve in the coming years. For example, the use of *learner model-based* adaptive learning and the application of machine learning to select learning activities has only just begun to influence online learning at large. Separately, social learning and collaboration have long been important to in-person settings. These are complicated to incorporate into the typically asynchronous adaptive learning systems, but research suggests that they could be very beneficial to some students. We encourage readers to reflect on their current teaching and explore ways that existing adaptive learning products could expand or improve their work. We also encourage the creation of new adaptive learning materials and continued innovation in this field.

The authors acknowledge support of the NASA Science Mission Directorate's Science Activation Program (Award #NNX16AD79G S01).

References

Blanchard, M. R., Southerland, S. A., Osborne, J. W., et al. 2010, SciEd, 94, 577
Ben-Naim, D. 2011, Doctoral dissertation, Univ. of New South Wales
Bloom, B. S. 1984, Educ Res, 13, 4
Brownell, S. E., Kloser, M. J., Fukami, T., & Shavelson, R. 2012, JCSTe, 41, 36
Bybee, R. W., Taylor, J. A., Gardner, A., et al. 2006, The BSCS 5E Instructional Model: Origins and Effectiveness (Colorado Springs, CO: BSCS), 88
Chi, M. T., Siler, S. A., Jeong, H., Yamauchi, T., & Hausmann, R. G. 2001, Cogn Sci, 25, 471

Cohen, P. A., Kulik, J. A., & Kulik, C. L. C. 1982, Am Educ Res J, 19, 237
Edelson, D. C. 2001, JRScT, 38, 355
Fosnot, C. T., & Perry, R. S. 2005, Constructivism: Theory, Perspectives, and Practice, ed. C. T. Fosnot (New York: Teachers College Press), 8
Hofstein, A., & Lunetta, V. N. 2004, SciEd, 88, 28
Horodyskyj, L. B., Mead, C., Belinson, Z., et al. 2018, AsBio, 18, 86
Kapur, M. 2008, Cogn Instr, 26, 379
Karplus, R., & Thier, H. 1967, A New Look at Elementary School Science; Science Curriculum Improvement Study (New Trends in Curriculum and Instruction Series) (Chicago, IL: Rand McNally)
Klahr, D., & Nigam, M. 2004, Psychol Sci, 15, 661
Kirschner, P. A., Sweller, J., & Clark, R. E. 2006, Educ Psychol, 41, 75
Koedinger, K. R., Kim, J., Jia, J. Z., McLaughlin, E. A., & Bier, N. L. 2015, Proc. Second ACM Conf. Learning@scale (New York: ACM), 111
Kulik, J. A., & Fletcher, J. D. 2016, Rev Educ Res, 86, 42
Lawson, A. E. 2010, Teaching Inquiry Science in Middle and Secondary Schools (Thousand Oaks, CA: Sage)
Ma, W., Adesope, O. O., Nesbit, J. C., & Liu, Q. 2014, J Educ Psychol, 106, 901
Mavroudi, A., Giannakos, M., & Krogstie, J. 2018, Interact Learn Environ, 26, 206
Mead, C., Buxner, S., Bruce, G., Taylor, W., Semken, S., & Anbar, A. D. 2019, JGeEd, 67, 131
Pearson, P. D., Moje, E., & Greenleaf, C. 2010, Sci, 328, 459
Settlage, J. 2007, JSTEd, 18, 461
Shute, V., & Towle, B. 2003, Educ Psychol, 38, 105
Shute, V. J., & Zapata-Rivera, D. 2007, Adaptive technologies, ETS Research Report Series, RR-07-05
Toven-Lindsey, B., Rhoads, R. A., & Lozano, J. B. 2015, Internet High Educ, 24, 1
U.S. Department of Education, 2013, Expanding Evidence Approaches for Learning in a Digital World (Washington, DC: U.S. Department of Education Office of Educational Technology)
VanLehn, K. 2011, Educ Psychol, 46, 197

Astronomy Education, Volume 2
Best practices for online learning environments
Chris Impey and Matthew Wenger

Chapter 8

Key Online Resources for Teaching Astronomy

Chris Impey and Andrew Fraknoi

The growth of the Internet has facilitated the easy availability of resources for teaching astronomy, particularly at the introductory level. This overview concentrates on those resources that are free or open access. Basic materials like textbooks, lab activities, and large numbers of astronomical images can be found, along with higher level items such as concept inventories and interactive instructional tools. Instructors can find teaching guides and tips for interdisciplinary approaches to astronomy. There is also a small but growing research literature on astronomy instruction to be found online. Taken together these resources are of great value to both novice and seasoned instructors.

Chapter Objectives

By the end of this chapter, readers will be able to:
- Describe the overall landscape for astronomy teaching resources online.
- Find free textbooks, lab activities, and interactive tools.
- Understand how astronomy instructors are using podcasts and citizen science.
- Find sky-viewing tools and image archives.

8.1 Introduction

The Internet has transformed the teaching of astronomy. Before 1995, instructors were mostly reliant on a printed textbook, 35 mm slides, and their own lecture notes. Depending on local resources, they might also be able to incorporate labs, hands-on activities, and the use of small telescopes. Now, they can choose from a wide, and occasionally bewildering, array of online resources to augment what they do in the classroom. The challenge is to find the tools that are either evaluated in peer-reviewed publications or have proven their efficacy in other ways.

On the other hand, experimentation in teaching is important, so the fact that any instructor or developer can put online the resources they have developed or found useful is a plus. This egalitarian aspect of the Internet weighs against the tendency of instructors to rely of a few authoritative sources of information. The other positive influence of the Internet is the tradition, strong among scientists and educators, to share materials freely and not put them behind paywalls.

Peer review of online resources is still quite rare, and since the unfortunate demise of the *Astronomy Education Review*, there is no single journal covering educational resources in our field. Astronomy instructors must keep current by networking with colleagues, by reading discussion forums and list-serves on astronomy education, and by attending education sessions at the meetings of the American Astronomical Society, the Astronomical Society of the Pacific, and the American Association of Physics Teachers. Suitable implementation of materials in the classroom is almost as important as the materials themselves, so instructors need to seek out best practices among their colleagues.

The list of online resources for teaching astronomy begins with digital versions of traditional tools like textbooks and lab activities. It includes learner-centered types of pedagogy like lecture tutorials, ranking tasks, and think–pair–share questions. There are also many forms of interactive tools like apps and simulations, extending to immersive online environments. Even the traditional planetarium experience has been taken online as an adjunct to students looking at the real night sky. Then there are the enormous collections of annotated astronomical images and short videos on astronomy topics freely available to everyone. Even research-level astronomy data can be used in the classroom by adapting "citizen science" projects that are generally intended for the public. Last, there are collections of resources and instructional guides that make connections between astronomy and other subjects that might especially appeal to non-science majors.

It is abundantly clear that the role of online resources has grown since the last survey of progress in introductory astronomy education (Waller & Slater 2011). In a review of such a dispersed and evolving online landscape, it is impossible to cover everything or be complete. Some instructors have put instructional tools online singly or in small numbers; these efforts are too numerous to represent properly. This article necessarily focuses on *collections* rather than individual items. There is also no full discussion here of the efficacy or research validity of the tools we list; interested readers should refer to the companion to this ebook, which is a practitioner's guide to research on astronomy instruction, rooted in research. The authors will endeavor to keep the links in this chapter current, and appreciate hearing from readers about any errors or major omissions.

The primary audience for the resources described in this article are instructors teaching an introductory astronomy course for non-science majors, and their students. However, many of the resources are also of interest to amateur astronomers and informal, adult (or lifelong) learners of astronomy. Most of them are freely available on the web.

8.1.1 Textbooks

Astronomy by Andrew Fraknoi, David Morrison, Sidney Wolff, et al. (OpenStax) (http://openstax.org/details/astronomy)
This free, open source textbook for teaching or learning introductory astronomy is a project of the non-profit OpenStax program at Rice University. The book, which can be used online or downloaded in several formats, was written and vetted with the assistance of about 70 astronomers (Fraknoi 2017). It's modular, so sections can be used independently. There is an Open Educational Resources Hub with the book, where ancillary materials by the authors or adopters are made available free of charge as well. The book has a host of links to effective videos and apps around the web, written and web-based resources that go with each chapter, profiles of a diverse group of astronomers, interdisciplinary connections and frequent touches of humor. Because the textbook is in digital form, it is updated annually, and can be corrected at any time. Students can download it or access it on their phones, tablets, or laptops wherever they are, without adding weight to their backpacks. About 300,000 students have used the book since it came out, distributed among about 900 institutions.

Teach Astronomy by Chris Impey et al. (www.teachastronomy.com)
This web site contains a range of tools for teachers and students of astronomy. The core is a comprehensive astronomy textbook, consisting of 520 articles, organized into 19 chapters. The articles derived from a traditional printed book authored by Chris Impey and William Hartmann "The Universe Revealed" (Impey & Hartmann 2000). After the intellectual property was released by the publisher, the textbook as recast as a set of articles, each of which can stand alone. The site also includes a curated set of 45,000 astronomy articles from Wikipedia, a series of 1200 short video clips covering most astronomical topics, and a unique clustering tool applied to the image collections from Astronomy Picture of the Day and Astropix and the collection of summaries of research articles at Astrobites. There is also an RSS feed of the most recent astronomy news stories from Science Daily. A cross-site search locates content of any media type with a keyword search. The articles can be explored in classic textbook order, or using a clustering tool called a Wikimap that shows closely related articles based on keyword overlap. For more details on the features of the site, and technical background on the clustering tool and the content database, see Impey et al. (2016).

Astronomy Notes by Nick Strobel (Bakersfield College) (https://www.astronomynotes.com)
This older site provides notes that are assembled into chapters and can serve as a free textbook. Some of the chapters have not been updated since the 2001–2012 time frame. The notes are more succinct than the above two books, but lots of clear diagrams and photos illustrate the text. There are areas of modern astronomy where these notes are not as thorough and up-to-date as instructors might want, but the brief coverage might appeal to students pressed for reading time.

8.1.2 Laboratory Activities

Compilation of Free Astronomy Lab Activities by Andrew Fraknoi (http://www.fraknoi. com/wp-content/uploads/2017/12/Laboratory-Activities-for-Astro-101.pdf)
This listing brings together both full collections of lab activities (mainly from university astronomy departments) and a selection of individual activities that are particularly useful. URLs are given to lead directly to each activity. Also, collections of high school level activities that have the potential to be used as part of lower-level college labs are listed. (The popular CLEA labs, Contemporary Laboratory Exercises in Astronomy, are included in this list.)

Astronomy Labs by Nate McCready and Emily Rice (http://www. mccradyricelabs.com/)
Among the many commercially published lab manuals that are available, this set of 40 lab exercises perhaps stands out because you can make your own selection and have your bookstore order a manual of only the activities you select (priced per activity). However, there are a good number of other printed lab manuals that can be found going in and out of print.

8.1.3 Compilations of Instructional Tools

ComPADRE (https://www.compadre.org/astronomy/)
ComPADRE is a digital library of educational resources in physics and astronomy intended for instructors and students The project was sponsored by the American Association of Physics Teachers and the American Astronomical Society, and has been supported by the National Science Foundation as a resource collection for the National Science Digital Library (NSDL). The collection is diverse, covering tutorials, activities, labs, simulations, animations, and papers and conference proceedings on physics and astronomy education research. The bulk of the materials relate to physics, but there are a substantial number of astronomy items, and the physics collections include many topics relevant to teaching introductory astronomy, such as radiation, light, atomic structure, and gravity (Deustua 2004).

This heterogeneous collection contains hundreds of astronomy items appropriate for a non-science major class. There was an original intention to have all materials be peer-reviewed, but that does not appear to have been carried out in practice. Note that the most recent addition to the astronomy collections was in 2016. No one from astronomy seems to be charged with maintaining or expanding this collection.

Merlot (https://www.merlot.org)
Merlot is a site for online college-level educational resources in a wide range of academic disciplines, supported by a number of statewide university systems and by volunteers around the country. As of 2018, about 800 astronomy examples of education materials were in the system. Unlike ComPADRE, this continues to be updated, although it is less specialized.

Center for Astronomy Education (https://astronomy101.jpl.nasa.gov/)
Hosted by NASA's Jet Propulsion Laboratory, the Center for Astronomy Education (CAE) features an extensive collection of resources for both instruction and assessment of introductory astronomy. The project is directed by Ed Prather at the University of Arizona. Based on over fifteen years of research on pedagogy and student learning, these materials have all been tested and validated (for more information, see Prather and Brissenden in the companion ebook).

The CAE web site hosts instructional strategy guides on using lecture tutorials and think–pair–share questions (see Wallace and Prather in the companion ebook). It also hosts images for use with lecture tutorials, and several new tutorials on radio astronomy and the detection of exoplanets; the actual lecture tutorials are published elsewhere (Prather et al. 2012). Over a thousand assessment questions on over 40 topics are provided for use with the pedagogical tools on the site. There are also nearly 50 sets of ranking tasks on topics in basic physics, positional astronomy and stellar astronomy. The site also has banks of multiple choice exam questions, and a full set of lecture slides for an introductory astronomy course.

CAE has been hosting workshops for instructors of astronomy and space science for over a decade. These include national workshops at biannual meetings of the American Astronomical Society, and regional and short course offerings that move around the country. There is also a CAE listserv, operated as a Google Group, with over 1200 people registered. The discussion thread is a rich source of ideas and tips for anyone teaching introductory astronomy.

Open Educational Resources Commons (https://www.oercommons.org/)
The OER Hub is a shared national library of educational resources for both the K-12 and college level. Within the hub, collections can be set up for specific topics or purposes. For Astro 101, there is a hub of shared resources set up through the OpenStax textbook, "Astronomy": https://www.oercommons.org/groups/openstax-astronomy/1283/?__hub_id=27.

Center for Astronomy and Physics Education Research (http://www.caperteam.com/)
Based at the University of Wyoming, the Center for Astronomy and Physics Education Research, or CAPER, is headed by Stephanie Slater and their web site lists a team of 15 scientists and researchers working on strategies and materials for the astronomy classroom. It has Android and Apple versions of the ABCD polling tool, an easy to use and low-tech version of classroom clickers (see Duncan in the companion ebook), and a series of Amazon links to books on astronomy instruction. CAPER sponsors a series of workshops and conferences, while discussions take place in a Facebook group with over 700 members.

8.1.4 Online Homework and Learning Management Systems

Mastering Astronomy (https://www.pearsonmylabandmastering.com/northamerica/masteringastronomy/)
Some publishers have tried to leverage their traditional astronomy textbooks by creating additional instructional resources that go beyond sets of multiple choice

quiz questions. The most successful and fully-featured of these experiments is *Mastering Astronomy* by the publisher Pearson, designed to be used with textbooks by Bennett et al. (2018) and Chaisson & McMillan (2017). Mastering Astronomy includes pre-lecture reading questions, quizzes, and interactive videos. The figures include narration and chances for students to pause and answer questions. There are also virtual labs and self-guided tutorials. Students have a customized path through the material, tailored to their responses and rate of progress, with learning analytics made available to the instructor. These capabilities come at a substantial cost to students. A subscription to Mastering Astronomy on its own costs $60. When that is bundled with access to the eText of the Bennett et al. book, for example, it costs $193. This including an online book, teaching materials, and integration with assignments, class rosters, and grades in the instructor's Learning Management System (LMS).

WebAssign (http://www.webassign.net/features/textbooks/astronomy_textbooks.html)
WebAssign is owned by the textbook publisher Cengage. It is not as extensive as *Mastering Astronomy*, but offers some similar features for the textbooks from its line.

Sapling Learning (https://macmillanlearning.com/catalog/page/Sapling/astronomy)
Sapling Learning is owned by the publisher Macmillan. These instructional resources can work with a number of different textbooks.

Expert TA (http://theexpertta.com/)
Expert TA began serving physics textbooks, but is now expanding to cover astronomy; a package to accompany the OpenStax textbook became available in 2020.

Canvas (https://www.canvaslms.com/)
Canvas is a generic Learning Management System, which entire institutions adopt, and instructors then program to work with their own classes and materials.

Other generic Learning Management Systems include *Blackboard, Moodle*, and *D2L Brightspace.*

8.1.5 Concept Inventories

A concept inventory is a research-based assessment instrument that probes a student's understanding of key concepts in a subject. Typically, it is administered with a carefully defined curriculum, and student learning is measured before and after the concept has been covered in class. Concept inventories were pioneered in physics (Hestenes et al. 1992), but they have since spread to astronomy and other subjects in science and beyond (Sands et al. 2018). Concept inventories have been developed on general space science and astronomy, positional astronomy, lunar phases, light and spectra, star properties, planet formation, and size, scale, and structure. For advice on administering concept inventories, see Madsen et al. (2017). Concept inventories have to be used carefully, as noted in an article on item response theory (Wallace & Bailey 2010).

Key astronomy concept inventories can be found at: https://www.physport.org/assessments/.

(Note that you must register as a faculty member at this site, kept by the American Association of Physics Teachers, to have access to the inventories.)

One of the earliest inventories, the Astronomy Diagnostic Test (ADT), can be found at: http://solar.physics.montana.edu/aae/adt/.

8.1.6 Short Videos

Videos can provide useful augmentation or enrichment for an astronomy class. The subject of astronomy has long been well served by long format (video from national media producers such as PBS/NOVA and National Geographic). But a relatively new phenomenon is short format video, often made by individual scientists, NASA or ESA missions, or educators, sometimes with inexpensive equipment. YouTube is the place to find many excellent videos on astronomy (and every other subject under the Sun). Started in 2005 and now operated by Google, YouTube is now the second most popular web site in the world. Over 400 hr of content are uploaded every minute, and over a billion hours of content is viewed every day (Zhou et al. 2016). A video web search for "astronomy" returns 2.9 million results, with about 5000 new videos added every day. A few noteworthy resources in this arena include:

The *Astronomy Crash Course* series, hosted by Phil Plait and created and distributes by PBS Digital Studios (https://www.pbs.org/show/crash-course-astronomy/).

There are 47 videos in the series, in the range 10–15 min long, with a total of 9.5 million views (Schmidt 2015).

Teach Astronomy project videos produced by the "Active Galactic Videos" team (https://www.youtube.com/ActiveGalacticVideos/).

This set of over 70 short videos covers everything from tours of major observatories to exploration of topics like the relativity, the HR Diagram, and the threat from asteroids. These videos resonate with students who are not science majors because they are conceived, scripted, acted in, and produced by undergraduate students.

Short Videos to Go with Each Part of Introductory Astronomy by Andrew Fraknoi (http://bit.ly/shortastronomyvideos)
This is a document collecting short (few minute) videos that go with each section of an Astro 101 course, with the URL and a short annotation given for each video. The videos come from a wide range of free sources, including space agencies and missions, public TV, observatories, and science museums. While this happens to be organized to go with the OpenStax *Astronomy* text, it could be used with any standard astronomy course sequence.

Other astronomy-specific channels include "Sci Show Space," (https://www.youtube.com/user/scishowspace) with a million subscribers, "HubbleCast" (https://www.youtube.com/playlist?list=PL206E0B9C3D070D40) and "Vintage Space" (https://www.youtube.com/channel/UCw95T_TgbGHhTml4xZ9yIqg), which is more about space flight than astronomy.

8.1.7 Interactive Tools

ClassAction (http://astro.unl.edu/classaction/)
Kevin Lee at the University of Nebraska has created a set of interactive materials on astronomy for use either at the introductory college level or the high school level. They are dynamic think–pair–share questions, with over 500 items across 22 topic modules. These resources can be imported into Powerpoint and there is a browser and module editor (for both PC and Mac) that allows instructors to customize their own modules. The web site also has instructions and comments on the pedagogy behind the materials. The menu at the top also provides links to the CAE ranking tasks, and a description of 20 virtual online labs that were created to go with two Brooks/Cole astronomy textbooks.

Nebraska Astronomy Applet Project (http://astro.unl.edu/naap/)
Also from Kevin Lee, the Nebraska Astronomy Applet Project (NAAP) is a set of 15 online lab modules for introductory astronomy. The labs are built around a set of simulators of physics and astronomy phenomena, with students able to set up initial conditions and vary parameter, so acting very much like scientists (see the article by Lee in the companion ebook). Each lab has background materials, and pretests and posttests that can be used to measure student learning. There is an instructor guide, a student guide, and a technical manual. Conversion of ten of the interactive simulations from Flash to HTML5 is underway, and it is planned to eventually convert all of them. As part of this conversion, David Helfand at Columbia University is preparing student-centered, discovery-based exercises to go with the simulations; another page by Geoff Mathews keeps track of which simulations from this and other project are available: https://foothill.edu/astronomy/astrosims.html.

PhET Interactive Simulations (https://phet.colorado.edu/)
Started in 2002 by Carl Weiman at the University of Colorado, the PhET project was the first to create simulations based on education research (Wieman et al. 2010), and the first to provide students with a game-like interface that encourages exploration and discovery. The app versions of the simulations work on both Android and iPhone devices, and they have been converted to HTML5 to work on every web platform and device. The web site has information on how to use the simulations most effectively and on the instructional scaffolding required to promote conceptual learning. Most of the PhET simulations are in physics, but there is overlap with material taught in an introductory astronomy course, with several dozen simulations on gravity, radiation, and stellar properties.

NASA's Eyes (https://eyes.nasa.gov/)
NASA's Eyes is an immersive app for desktop PCs and Macs, and for mobile devices. It's the work of the Visualization Technology Applications and Development Team at the Jet Propulsion Laboratory. The app is a technologically sophisticated tool for embedding NASA's planetary data assets in an environment of data visualization where the user has flexibility in exploring the data. The "Eyes on the Earth" module lets a user monitor the planet's vital signs, trace the movement of water around the planet, and interact with a temperature map of the planet. The "Eyes on the Solar System" module allows a user to inspect planet and moon surfaces with mapping data from NASA missions, and lets a user recreate solar system exploration with missions from 1950 to 2050. The "Eyes on Exoplanets" module renders 3D space and populates it with over 1000 confirmed exoplanets. Users can inspect physical properties of the exoplanets, make comparisons between the exoplanet systems and solar system geometry, and overlay their habitable zones. The app is relatively new, so no resources for instructors are available yet.

8.1.8 Sky-viewing Tools

The number of programs and apps to help students learn and explore the sky has grown...well... astronomically in recent years, with new and powerful versions coming out regularly. This arena really deserves a review article of its own, by someone more current in the field than the present two authors. Here we just mention a few tools that have proven especially useful in educational settings. There are regrettably few comparative reviews of these tools in the literature by people knowledgeable in astronomy and education. One useful article is Young (2015).

WorldWide Telescope
The Internet enables views of the night sky to be integrated with images from ground- and space-based telescope, and these capabilities have great potential for teaching astronomy. The exemplar of this is WorldWide Telescope (WWT), an open source collection of applications and data, hosted on GitHub, with the data available in the cloud (http://www.worldwidetelescope.org/webclient/). Originally developed by Microsoft Research and available only as a Windows application, WorldWide Telescope now has a web client so it can be used in a browser on any desktop computer or handheld device. The project is managed by the American Astronomical Society. It realizes the long-held vision of an open source, virtual observatory (Gray & Szalay 2002). WWT has been downloaded by over 10 million active users.

WWT can display terrestrial, planetary, or astronomical data, with navigation by panning or zooming through a 3D universe. Third-party data can be imported if it is in WWT format. The multi-wavelength universe is available to view; current collections include data from Hubble, Chandra, Spitzer, 2MASS, and radio telescopes. There are all-sky data sets from the Digitized Sky Survey, IRAS, Fermi, GALEX, the VLA, and the WMAP and Planck microwave satellites. Full maps of the Earth and Mars can be explored. WWT can function as a planetarium viewing tool, with constellation maps and full control of time and location on the Earth. The web client can even be used to drive full-dome projection in a real planetarium. For more details on these capabilities, see Goodman et al. (2011) and the article by Udomprasert in the companion ebook.

The capabilities of WorldWide Telescope can seem overwhelming to an astronomy instructor, but the project already provides over 50 examples of "tours" that instructors can use in the classroom to give students a sense of the richness of the night sky. A tour has a narrative and film-like quality that makes it very engaging. Instructions online let any teacher create their own tours, and share them with the user population. There is a WWT Ambassador program run out of Harvard University that trains high school teachers in the use of the tool and offers workshops at AAS meetings.

Google Sky (https://www.google.com/sky/)
As a counterpoint to its comprehensive maps of the Earth, Google has for over a decade made available sky maps. Like the WorldWide Telescope, image archives of the major observatories are included. Zoomable maps of the Moon and Mars are also available, and the Moon resources include some 3D models and 360° panoramas.

Stellarium (http://stellarium.org/)
There are a number or free or open source planetarium software packages that could be used in a lab or classroom setting for introductory astronomy. The most prominent is the free open source *Stellarium*, which is supported on all major operating systems. In addition to visible sky objects, and constellation maps from ten cultures, Stellarium has 600,000 stars from the Hipparcos and Tycho-2 catalogs, and the ability to import 200 million more. There is a mobile version for Symbian, Android, and iOS (Hughes 2008).

Other Apps
Other well-reviewed free programs are *Celestia* (https://celestia.space/), *SkyChart* (https://www.ap-i.net/skychart/en/start), and *Aladin* (http://aladin.u-strasbg.fr/). For examples of how to use these tools in the classroom, see Persson and Eriksson (2016). The most widely-used commercial planetarium software is *Starry Night College*, which has lesson plans, pre- and post-assessment resources, and interactive student exercises, but charges significant fees for users.

NASA MicroObservatory (http://mo-www.cfa.harvard.edu/MicroObservatory/)
The MicroObservatory is a network of robotically-controlled, 6 inch telescopes that take images for educational purposes. The project is funded by NASA and it was developed by scientists and educators at the Harvard–Smithsonian Center for Astrophysics (Gould et al. 2006). The network has mostly been used at the high school level, but it provides an excellent way for non-science majors in college to get a taste of real observing by gathering their own data. In the "Other Worlds" project, middle and high school students from seven states used the network to detect

transiting exoplanets. Typically, 50–60 bright objects can be observed by the small telescopes, including planets, star clusters, and a few nebulae and galaxies. Students select their object, the filter, and a field of view, and submit a request. Data is typically taken within a week by one of the telescopes in the network.

8.1.9 Citizen Science

Citizen science is a phenomenon that is a direct result and beneficiary of the spread of Internet access around the world. Millions of volunteers are working of thousands of projects across all fields of science, without any formal training. Some of them work on research-level data sets; others are crowd-sourcing measurements of the natural world (Bonney et al. 2014).

To find citizen science projects, you can use the database at: https://scistarter.com/.

Galaxy Zoo (https://data.galaxyzoo.org/)
The archetypal citizen science project in astronomy is Galaxy Zoo. The original project ran from 2007 July to 2009 February. Volunteers classified galaxies from the Sloan Digital Sky Survey into six categories according to their morphology. It was stunningly successful; non-scientists classified 900,000 galaxies with a reliability not very different from that of trained professionals (Lintott et al. 2008, 2010). There have been 15 different projects in Galaxy Zoo. They include work on detailed morphological classifications, looking at mergers, looking for supernovae, and looking at deep imaging data from the Hubble Space Telescope's CANDELS survey.

Galaxy Zoo expanded a few years ago into Zooniverse (https://www.zooniverse.org/). Zooniverse is operated by the Citizen Science Alliance as an umbrella for many citizen science projects. Citizen scientists are working or art and archaeology, weather data and animal classification. The site has over a million registered volunteers, and their efforts have contributed to over a hundred research papers. A range of astronomical projects are underway, including looking for coronal mass ejections in the Sun (Solar Stormwatch), detecting bubbles in the interstellar medium (The Milky Way Project), using light curves to detect extrasolar planets (Planet Hunters), analyzing images of Mars (Planet Four), looking for stars where planets are forming (Disk Detective), and analyzing time-lapse images to find undiscovered asteroids (Asteroid Hunter).

Other Projects
There are other citizen science project organized around NASA data. CosmoQuest enlists the public to help make maps and identify features on the Moon, Mars, Mercury, and Vesta (https://cosmoquest.org/x/). Astronomy Rewind is using volunteers to map and digitize sky images from old astronomical journals.

Marshall et al. (2015) have described the landscape of astronomy citizen science. There are projects that crowd-source desktop computing resources to discover artificial signals from space (SETI@Home) and gravitational waves (Einstein@Home), but since they don't actively engage users in the task, they are of limited instructional value.

Using citizen science in the astronomy classroom takes careful planning, to make sure there is sufficient training and motivation for the work. Students taking astronomy for a General Education requirement are driven largely by grades and are not generally very committed to the subject matter. The science-curious public have a high degree of self-motivation that lets them persist with long and sometimes tedious tasks (Raddick et al. 2013). The California Academy of Sciences has published a useful guide for integrating citizen science projects into classroom curricula. Although it is aimed at high schools, the same principles apply for college level introductory astronomy (California Academy of Sciences 2015: https://www.calacademy.org/educators/citizen-science-toolkit).

8.1.10 Interdisciplinary Approaches to Teaching Astro 101

Since most students taking this course are not science majors, it can be effective pedagogy to show them connections between astronomy and the fields they are more familiar with, such as music, poetry, science fiction, drama, etc. This could be a brief aside in class, or part of homework or extra-credit assignments. Also, students may get interested in (and even involved with) social issues connected with astronomy, such as light pollution or the role of women in society. We note that there is a national push to integrate science and the arts in higher education (National Academies of Sciences, Engineering, and Medicine 2018). Here are some topical resource guides to help you and your students explore these topics.

- *An Annotated Guide to Plays (and a few Films) about Astronomers* (http://bit.ly/astroplays).
- *Science Fiction Stories Based on Reasonable Astronomy* (http://bit.ly/astroscifi).
- *Music Inspired by Astronomy: A Resource Guide Organized by Topic* (http://bit.ly/astronomymusic).
- *Astronomy and Poetry: A Resource Guide* (http://doi.org/10.3847/AER2001009).

The Inspiration of Astronomical Phenomena (https://insap.org/).
These are regular international conferences connecting astronomy with the humanities and the arts. Books of collected papers are published after each conference.

Donald Olson: Connecting Astronomy with History and Art (http://donolson.wp.txstate.edu/).
Prof. Donald Olson of Texas State University works with students to solve mysteries or puzzles connecting historical events or works of art with astronomical phenomena. He has published two Celestial Sleuth books on his investigations (Olson 2014, 2018).

Social Issues:
- *A Resource Guide on Light Pollution and Radio Interference* (http://www.fraknoi.com/wp-content/uploads/2018/03/Light-Pollution-Bibliography-2018.pdf).
- *A Resource Guide on the Role of, and Challenges for, Women in Astronomy* (http://bit.ly/astronomywomen).
- *Multi-cultural Astronomy:* The astronomy of non-western cultures (http://bit.ly/astrocultures). See also, Black Lives in Astronomy (http://bit.ly/blackastro).
- *Web Resources for Debunking Astronomical Pseudo-science*—such as Moon-landing denial, young-Earth creationism, and UFO's interpreted as spaceships (http://bit.ly/pseudoastro).

8.1.11 Podcasts

Podcasts are a series of episodes of digital audio or video files that a user can listen to, subscribe to, and download. The format has been around for almost twenty years, but it is currently surging in popularity due to standard digital formats and the widespread use of Internet-enabled phones. Podcasts are useful resources for teaching if instructors can use them for enrichment or for having students write a short paper or commentary. Episodes that work in that capacity are stories of recent discoveries, descriptions of exotic or "hot" astronomical objects, or interviews with prominent astronomers.

The highest profile astronomy podcast is *StarTalk Radio*, hosted by Neil deGrasse Tyson (https://www.startalkradio.net/). Other long-term efforts include *Astronomy Cast*, hosted by Fraser Cain and Pamela Gay (http://www.astronomycast.com/), and a compilation of recent space science news with *NASACast Audio* (https://www.nasa.gov/podcasts).

One of the longest running podcasts, with daily episodes, is *365 Days of Astronomy* (https://cosmoquest.org/x/365daysofastronomy/). *Naked Astronomy*, produced by the BBC, is part of the Naked Scientists franchise. (https://www.thenakedscientists.com/podcasts/astronomy-podcasts).

There are also nationally syndicated science podcasts that sometimes cover astronomy topics: *Science Friday, Radiolab, Quirks and Quarks, Big Picture Science* (from the SETI Institute) and *60-Second Science.*

8.1.12 Image Collections

Resource Guide to Finding Astronomical Images for Teaching by Andrew Fraknoi (http://www.fraknoi.com/wp-content/uploads/2017/12/Astronomical-Image-Sources-2018.pdf)
One of the best parts of teaching astronomy these days is the easy availability of beautiful and instructive images from the world's observatories and space missions.

This annotated guide lists and gives links to 23 online sources for astronomical images that are useful for the Astro 101 instructor. We list three image sites that are perhaps most general below, but encourage you to explore the others as well.

Astronomy Picture of the Day (https://apod.nasa.gov/apod/astropix.html)
The iconic image web site Astronomy Picture of the Day (APOD) started at the same time as the Internet, in 1995. Its first post had 14 views, and by 2012 it had reached one billion views. The site is hosted by NASA and Michigan Technological University, and the images it presents have played an important role in anchoring astronomy in the consciousness of the public (Bonnell & Nemiroff 2006). With over 8000 images plus explanatory captions, it is one of largest repositories of astronomical imagery on the Internet. Bob Nemiroff, one of the co-creators of the site, has posted information on how to use APOD in the classroom, including slides from an Astro 101 class he has taught using only Wikipedia and APOD rather than any textbook (https://apod.nasa.gov/apod/lib/apodclass.html).

Astropix (https://www.ipac.caltech.edu/outreach/project/astropix)
Hosted by the Infrared Processing and Analysis Center (IPAC) and sponsored by NASA, Astropix is an archive of over 7000 images from the world's major observatories. In particular, it has assets from the Hubble, Spitzer, Chandra, Spitzer, GALEX, Herschel, Planck, and WISE telescopes in space, and from ESO's ground-based telescopes. Astropix adds value to images by aggregating all the relevant contextual information about an image, such as the source, sky position, field of view, wavelength of observation, and color map. This metadata is part of a worldwide standard called Astronomical Visualization Metadata (AVM). For an instructional use, it is easy to compare images of the same object taken across the electromagnetic spectrum.

Planetary Photojournal (https://photojournal.jpl.nasa.gov)
Supported by a team at JPL, this site collects and organizes the images of the solar system from NASA missions, currently about 22,000. You can search by world or by feature name, and the latest images are always available in a separate section. We wish the site would also incorporate planetary images from ESA, JAXA and other space agencies.

ESO Image Archive (https://www.eso.org/gallery/)
The largest source of astronomical images outside the United States is the archive of the European Southern Observatory (ESO). There are over 12,000 images on their site, plus over 3300 short videos and animations, organized by subject area of astronomy, and free to use for educational purposes.

8.1.13 Guides for New Instructors or TAs

Bruning, D. "Teaching Your First Astro 101 Course": (https://www.oercommons.org/courses/teaching-your-first-astro-101-course?__hub_id=27).

This long packet includes handy sample syllabi, lists of things to think about, course policies and assessment tools.

O'Connell, R. "Facts of Life for New Teachers in the Astronomy Non-majors Curriculum": (http://doi.org/10.3847/AER2007010).
A long guide for TA's and others new to teaching, from an experienced professor and his colleagues at the University of Virginia, published in *Astronomy Education Review*.

Tips for New TA's from the University of Washington Astronomy Department Teaching Clearinghouse (https://sites.google.com/a/uw.edu/introductory-astronomy-clearinghouse/teaching-methods/tips-for-new-tas).
Includes a wealth of practical information, from first-day ice-breakers, to sample handouts, to reasonable expectations, to best practices.

International Teaching Assistant's Guide from Vanderbilt University (https://cft.vanderbilt.edu/guides-sub-pages/itas/).
A clear introduction to roles, etiquette, expectations for students from other countries who are TA's at American universities.

Teaching Assistant Training Materials from the Science Education Initiative (https://www.colorado.edu/sei/fac-resources/other.html#teaching).
Good resources from the project at the University of Colorado and the University of British Columbia, led by physics Nobel Laureate Carl Wieman.

Goals for Astronomy 101 (http://doi.org/10.3847/AER2003016).
Workshops on the teaching of introductory astronomy were held in 2001 for astronomy department chairs and other leaders at selected major research universities to examine the goals for "Astro 101" courses. The full report, authored by Bruce Partridge and George Greenstein, was published in this paper in *Astronomy Education Review*.

8.1.14 Databases of Astronomy Education Research Articles

ISTAR (http://istardb.org/)
This database of astronomy education research papers and articles available on the Web contains about 2000 entries as of summer 2020.

Subject Index to Papers in Astronomy Education Review (2001–2013) (https://aas.org/teach/subject-index-papers-astronomy-education-review-2001-2013)
Astronomy Education Review was the first online journal devoted to the field, published by NOAO and then the AAS. This subject index allows you to consult any paper in the perpetual archive of the journal.

SABER (http://stellared.org/saber/biblio.pdf)
SABER was an attempt to set up a database of astronomy education research papers, and many from 1973 to 2009 were included. The effort is defunct, but a 38 page PDF printout of the database, organized by journal, is still available.

8.1.15 Blogs and Social Media Vehicles

A Listing of Blogs by Astronomers at *AstroBetter* (http://www.astrobetter.com/wiki/Blogs+by+Astronomers).
AstroBetter is a rich resource for younger astronomers wanting to improve many aspects of their work. This wiki page of bloggers is organized into seven categories.

CAPER Astronomy Faculty Lounge (https://astronomyfacultylounge.wordpress.com/).
A blog specifically devoted to Astro 101 issues.

Astronomy Education Facebook Group (https://www.facebook.com/groups/astronomyeducation/).
Administered by Tim and Stephanie Slater, this group has about 800 members and discusses topics of interest to both Astro 101 and K-12 instructors.

Center for Astronomy Education (https://www.facebook.com/groups/astronomyeducation/).
Administered by Ed Prather and members of his group at University of Arizona, this group has about 700 members and deals mostly with Astro 101 issues.

8.1.16 Miscellaneous Resources

AstroBetter Wiki Page on Resources for Astro 101 Teaching (http://www.astrobetter.com/wiki/tiki-index.php?page=Astro+101+Resources).
This communally supported page has a good listing of resources, including some not listed in this paper.

ConcepTests in Astronomy (http://hea-www.harvard.edu/~pgreen/educ/concep/Contents.html).
Now somewhat dated, this page contains a large shared database of multiple choice questions collected by Paul Green at Harvard.

Student Learning Objectives
(https://www.researchgate.net/publication/268277729_Writing_and_Assessing_Student_Learning_Objectives_Tips_Techniques_and_What_Our_Community_Needs).
Most Astro 101 instructors are now being asked to develop formal student learning objectives for their courses. This important task is being undertaken separately at each institution with no national coordination or assistance. This document is a report from the first conference discussion of SLOs and contains a resource guide to help instructors see what others are doing.

Astronomy Course Syllabi from the University of California, Santa Cruz (http://astro.ucsc.edu/courses/ugcoursewebpages.html).
Here you have access to a wide range of introductory courses offered in the last two decades by a wide range of instructors.

A very detailed Astro 101 syllabus by Dr Ana Larsen at the University of Washington can be seen here: https://canvas.uw.edu/courses/1063109/assignments/syllabus.

Guides for Students on How to Study for Astro 101:
Bennett, J. Hints on how to Succeed in College Classes (http://www.jeffreybennett.com/pdf/How_to_Succeed_general.pdf).
Fraknoi, A., et al. How to Study for an Introductory Astronomy Class (https://courses.lumenlearning.com/astronomy/chapter/how-to-study-for-an-introductory-astronomy-class/).
Seeds, M. & Geller, H. How to Study Astronomy (http://physics.gmu.edu/~hgeller/studyas2.html).
Virginia Tech Study Skills Self-Help for Students (http://www.ucc.vt.edu/stdysk/stdyhlp.html).

References

Bennett, J. O., Donahue, M., Schneider, N., & Voit, M. 2018, The Essential Cosmic Perspective (New York: Pearson)
Bonnell, J. T., & Nemiroff, R. J. 2006, Astronomy: 365 Days (New York: Harry N. Abrams)
Bonney, R., Shirk, J. L., Phillips, T. B., et al. 2014, Sci, 343, 1436
California Academy of Sciences 2015, Citizen Science Toolkit (San Francisco, CA: California Academy of Sciences) https://www.calacademy.org/educators/citizen-science-toolkit
Chaisson, E., & McMillan, S. 2017, Astronomy Today: Stars & Galaxies (9th ed; New York: Pearson)
Deustua, S. E. 2004, Mercu, 33, 19
Fraknoi, A. 2017, PhTea, 55, 502
Goodman, A., Fay, J., Muench, A., Pepe, A., Udomprasert, P, & Wong, C. 2011, in ASP Conf. Ser. 461, ed. P. Ballester (San Francisco, CA: ASP), 267
Gould, R., Dussault, M., & Sadler, P. 2006, AEdRv, 5, 127
Gray, J., & Szalay, A. 2002, Commun ACM, 45, 50
Hestenes, D., Wells, M., & Swackhamer, G. 1992, PhTea, 30, 141
Hughes, S. W. 2008, SEN, 57, 83
Impey, C. D., & Hartmann, W. H. 2000, The Universe Revealed (Pacific Grove, CA: Brooks-Cole)
Impey, C. D., Hardegree-Ullman, K. K., Patikkal, A., & Austin, C. L. 2016, IJIER, 4, 117
Lintott, C. J., Schawinski, K., Slosar, A., et al. 2008, MNRAS, 389, 1179
Lintott, C., Schawinski, K., Bamford, S., et al. 2010, MNRAS, 410, 166
Madsen, A., McKagan, S. B., & Sayre, E. C. 2017, PhTea, 55, 530
Marshall, P. J., Lintott, C. J., & Fletcher, L. N. 2015, ARA&A, 53, 247
National Academies of Sciences, Engineering, and Medicine 2018, in The Integration of the Humanities and Arts with Sciences, Engineering, and Medicine in Higher Education: Branches from the Same Tree, ed. D. Skorton, & A. Bear (Washington, DC: National Academies Press)
Olson, D. 2014, Celestial Sleuth: Using Astronomy to Solve Mysteries in History, Art and Literature (New York: Springer)
Olson, D. 2018, Further Adventures of a Celestial Sleuth (New York: Springer)

Persson, J. R., & Eriksson, U. 2016, PhyEd, 51, 025004
Prather, E.E., Slater, T.F., Adams, J.P., & Brissenden, G. 2012, Lecture-tutorials for Introductory Astronomy (3rd ed.; New York: Pearson)
Raddick, M. J., Bracey, G., Gay, P. L., et al. 2013, AEdRv, 12, 010
Sands, D., Parker, M., Hedgeland, H., Jordan, S., & Galloway, R. 2018, High Educ Pedagog, 3, 173
Schmidt, J. T. 2015, J Gilded Age and Progress Era, 14, 284
Wallace, C. S., & Bailey, J. M. 2010, AEdRv, 9, 010116
Waller, W. H., & Slater, T. F. 2011, JGeEd, 59, 176
Wieman, C. E., Adams, W. K., Loeblein, P., & Perkins, K. K. 2010, PhTea, 48, 225
Young, M. 2015, S&T, 129, 68
Zhou, R., Khemmarat, S., Gao, L., Wan, J., & Zhang, J. 2016, Multimed Tools Appl, 75, 6035

Chapter 9

Epilogue: Lessons Learned from Transitioning to Online Learning During Spring 2020 During COVID

Sanlyn Buxner, Nicole Gugliucci, Carl Ferkinhoff and Brian Jackson

In the Spring of 2020, as university and college campuses closed due to concerns of COVID-19 spreading across the world, students and instructors struggled to adjust to a new reality that involved staying at home, relying heavily on online course management systems in new ways, and finding new ways to engage with others despite being remote. This involved too many "new" things to take on all at once, and every mistake that could be made was made. With this epilogue, we hope you can learn from past mistakes and our lessons learned, which will continue to increase as we all move forward. As we advance beyond this frantic transition to remote learning, we reflect on how to move forward and continue to plan for future instruction in an increasingly uncertain world for higher education.

As astronomy instructors across the world reached out for advice, lessons, and resources, we found that all our circumstances were very similar. In the following narratives, you can read the experiences of faculty whose institutions span the University of Arizona—a public research institution with an undergraduate population of over 45,000, a large graduate teaching contingent, and introductory classes with hundreds of students; Boise State University with an undergraduate population of around 22,000; Winona State University with an undergraduate population around 8000; and Saint Anselm College, a small, liberal arts, Benedictine college with just over 2000 undergraduate students. Despite our different teaching assignments in the Spring of 2020, we all had to adapt to remote learning with little to no notice. May you learn from what we experienced.

9.1 Challenges and Considerations

At the University of Arizona, instructors had many choices in adjusting to online learning. Some instructors chose to convert their in-person classes to synchronous courses held over Zoom, the supported campus video-conference software, with accompanying asynchronous labs, pre-recorded lectures, and other assignments outside of class. Some instructors were able to set up group work during class while some required collaboration outside of class. Other courses transitioned to entirely asynchronous environments where students were required to complete videos, readings, and assignments outside of class time. Students across campus reported confusion as they had to navigate up to seven courses with instructors who had each chosen different fully course configurations. Additionally, instructors struggled with delivering courses fully online and using all the tools available. Constant student communication became a priority as students returned from Spring break to an entirely different learning experience. Additionally, students were given the choice to switch from regular course grades to pass/fail grades, often resulting in decreased motivation for participating in the second half of the semester. As the first few weeks unfolded, it became apparent that our traditional students had other priorities such as family, jobs, and their physical and mental health to contend with. All of these became even harder for non-traditional students. Challenges with online tools emerged, including being "Zoom-bombed" by unwelcome content during class. Students reported depression, isolation, and stress since they needed to complete courses to graduate.

9.2 Resources

In the first few weeks of the national shutdown, resources were shared across listservs to support instruction that remain relevant in future planning. Some were for university teaching in general such as the Chronicle of Higher Education (https://www.chronicle.com/article/how-to-be-a-better-online-teacher/), community based resources became available such as a centralized list of where to find your own university resources for teaching, https://docs.google.com/spreadsheets/d/1VT9oiNYPyiEsGHBoDKlwLlWAsWP58sGV7A3oIuEUG3k/edit#gid=1552188977 as well as crowd sourced resources for all aspects of teaching online across different disciplines.

https://docs.google.com/spreadsheets/d/1sVI7O3FrFzQFyeiYUQzsGvJWnWMxPY-NedI1kio_OpY/edit#gid=499998950. Other resources became available such as this one on humanizing online teaching https://digitalcommons.stmarys-ca.edu/school-education-faculty-works/1805/ as well as important resources to help instructors become more cognizant of "inclusion, equity, and access while teaching remotely" https://cte.rice.edu/blogarchive/2020/3/13/inclusion-equity-and-access-while-teaching-remotely. Additionally, broad resources for teaching online https://acue.org/online-teaching-toolkit/ and a network of instructional designers became available for instructors who were starting from scratch in online teaching https://www.idernetwork.com/.

Specifically for teaching astronomy, the astrolrner (astrolrner@googlegroups.com) listserve community provided many resources for teaching astronomy online. Dr Angelle Tanner from Mississippi State University started compiling resources for

teaching astronomy https://docs.google.com/document/d/1OsuFfqrlYa9diTJyTTzw ADrlk-4Ba7dghuMzw90p-og/edit?fbclid=IwAR0BBNFQijDTTGOP90U_e7h7Di O1Jo_G_7TFFuxjPwpwSzq2dt2A1yDIeBc and the AAS provided a page of curated resources that was a compilation of what had been shared by the astronomy teaching community and provided videos related to teaching.
https://docs.google.com/document/d/1UPZVyq2U7GTC5I7R1TI_rsRDwu4FX t9aukGYXIeq60Q/edit?usp=sharing.

These resources continue to be updated as instructors develop new resources for upcoming semesters.

9.3 Success Stories of Transition
9.3.1 Winona State—Carl Ferkinhoff

At Winona State, due to the technology resources available, I was able to move my courses online essentially unmodified, with the exception of moving to online simulation-based labs. I was able to continue to employ a flipped lecture format with students watching and responding to a pre-lecture video and questions and then participating in a student-centered lecture with think–pair–share questions and collaborative group work via Zoom. Students collaborated with partners via Zoom's breakout room feature and voted on think–pair–share responses via Zoom's polling system. As the questions were already in the lecture slides, with responses labeled as A, B, C, etc., it was easy to use a generic poll in Zoom that could simply be reloaded for each question we discussed. During group practice, students would use the whiteboard and annotate features in Zoom to collaboratively solve problems. If students have access to a tablet with stylus or drawing tablet and digitizer, the experience of using the whiteboard and annotate features in Zoom is nearly indistinguishable from working in-person. If students just have their finger or mouse, the experience isn't as good, but it still enables a good learning environment. Overall, I was surprised at how authentic to a normal in-person class the experience was, and on course evaluations, students greatly appreciated how similar the online experience was to the in-person experience. It was possible to have a quality online experience!

Some of this success arose from significant time learning about the learning management system and Zoom, as well as working with students to be comfortable with the technology. For instance, once students are moved into Zoom breakout rooms, they can no longer see the screen the host is sharing, which prevents students from seeing slides shared in the main Zoom room. Additionally, if a group wants to share their work on a whiteboard outside of the breakout room, they will need to save it before they leave the breakout room. I practiced using the technology myself and then practiced with my students.

One place that I saw difficulty was for students used to turning in handwritten work. Students without document scanners took photos of their work using their phone, but some phones use proprietary image files. If students didn't know how to verify their images uploaded to their learning management system, I was faced trying to grade assignments I could not view. After some initial struggle, I asked students to use a photo scanning app like Microsoft Lens that lets one compile multiple images into a single PDF like a scanned document.

At Winona State, in addition to trimming course content and changing delivery, some faculty chose to modify the course expectations. Once we moved online, I made much of the work optional. I assigned students a minimum grade based solely on their work prior to our transition to online. All work then after the transitions could only improve their grade. Further, students were required to only complete their exams and maintain some participation in the course. This requirement could be completing the homework assignments, participating in lectures or pre-lectures, or even just watching the recorded Zoom lectures. This approach allowed students to pick the level of participation based on their needs and personal situations. In reality, most students continued to participate regularly, but it removed some of their stress during a very stressful time, while acknowledging that an online course was not what they originally signed up for. Moving into the fall, since my students will know the format of the course ahead of time, this same level of flexibility will likely not be offered, yet some will still be needed. With an open campus—even with mask wearing, largely online or hybrid classes, and social distancing—some students and faculty will probably need to be quarantined during the semester.

9.3.2 Saint Anselm College—Nicole Gugliucci

At Saint Anselm College, my astronomy course had to shift from a "just-in-time teaching" approach pre-Spring break, in which students were able to reflect on the content and impact what happened during each class by giving feedback on the pre-class readings and small group work in-class that had taken advantage of learning communities. Luckily, we had an early spring break, and we had an opportunity to prepare students for the shutdown by doing a Zoom test before students were sent home. Although we had some notice and amazing support from our IT department, my experience teaching asynchronous was lacking and the resulting part of the semester was not ideal. I decided to rethink the entire second half of the course.

The structure of the remote classes went as follows: students still had a reading before we met and the same just-in-time reading assignment to complete beforehand. I recorded my mini-lectures so they could also view those before our usual class time. Then, we met over Zoom for a short discussion of the topic where I posed several questions and took notes on screen while they brainstormed answers. The class discussion, however, was not required, since it was clear to me that requiring synchronous meetings during a global pandemic was not going to be equitable or fair. Students, therefore, also submitted answers to discussion questions via Canvas's discussion board, which were graded for participation. Typically, the Zoom discussion built upon the posts that had already been written on the board. Our Zoom sessions were also recorded to allow anyone who was not there to view them.

Labs had to be adapted. For one such lab, I had the students explore and participate in an online citizen science project, in this case, Planet Four (https://www.zooniverse.org/projects/mschwamb/planet-four). This was very different from previous labs in that they could delve into the messier side of the process of science as they struggled to distinguish between "blotches" and "fans" in the Martian polar region and reflect on the benefits and potential pitfalls of such crowd-sourced science analysis. The course dedicated more time to facilitate writing assignments than in previous semesters. Although very few students elected to meet in-person over Zoom, I spent hours commenting on drafts in Word or Google docs. In fact, removing the option to hand in a physical copy of a draft made it even easier to suggest revisions and make comments on student drafts. Canvas also has a peer-review function so that, instead of the planned in-person peer reviews, students could comment on each other's drafts in the same way that I could, and I was able to see the reviewer comments.

Overall, despite the changes, I felt I had achieved the main learning goals of the course, even though I had cut a good bit of content that I had covered in previous semesters. Although the college was offering a generous Pass/No Pass option to all students for the semester, I didn't want students to agonize over their GPA, and so I relaxed my essay-grading standards in addition to cutting several assignments. I didn't bother with timed exams or online proctoring services. I kept my essay-format exams and made the questions a bit more open-ended in exchange for making them open-book and untimed. I also decided to replace the traditional final for a reflection piece that asked them to describe an idea, attitude, opinion, or belief that was challenged, changed, and/or reinforced over the course of this semester using seven guiding questions, inspired by a blog post by Francis Su, a professor of mathematics at Harvey Mudd College (https://www.francissu.com/post/7-exam-questions-for-a-pandemic-or-any-other-time). They had the option to submit via various formats: essay, PowerPoint presentation, letter to a friend, podcast, or video. Through this assignment, I discovered that the course had actually changed several students' minds about the importance of science, spaceflight, and the search for life in the universe.

9.4 Final Thoughts

By the end of the Spring 2020 semester, we were all burnt out but grateful we had made it to the end. Some of our universities chose to forgo formal teaching evaluations but we did receive notes from students thanking us for making courses enjoyable and less stressful during the quarantine when everything else seemed to be in flux. For those that did keep formal evaluations, scores did not suffer as one might have expected due to the perceived chaos.

The "impossible semester" reinforced several essential aspects of teaching. First, it was important to come back to focus on the learning goals of the course and build adapted lessons around those. Second, awareness of the students' situations and struggles was essential. Even though it was hard to do this in larger courses, it was easy to look and see who was not logging into the course management system or submitting assignments to prioritize students who needed additional attention. Frequent emails to remind students of due dates and strategies helped them navigate the stormy semester. Increased flexibility became ways to support students struggling with many competing priorities. Asking about students' access to and comfort with technology needed to complete the course became ways to understand hardships and let them provide additional information about their situation (e.g., starting with a simple poll, inspired by Julia Kamenetzky, an assistant professor of physics at Westminster College). Some of us found that forgoing timed exams and other traditional assessments forced us to design summative assessments that were higher on the pyramid of Bloom's taxonomy and helped alleviate stress for both us and our students. Assessments continue to be a struggle as we move forward.

No matter how we plan our courses for the future, we realize that we need to be prepared to go online with little notice again. We have found that the virtual community of astronomy instructors has been incredibly helpful in figuring out how to do this and continues to do so. There are many challenges that lie ahead for our courses that will have at least some components virtually and need as much flexibility as possible. We hope that this volume provides support in your own transition.

www.ingramcontent.com/pod-product-compliance
Ingram Content Group UK Ltd.
Pitfield, Milton Keynes, MK11 3LW, UK
UKHW061158160426
5217IPUK00049B/141